江南园林艺术手绘图志

Southern Chinese Garden Art Drawings

吴肇钊　著

中国建筑工业出版社

图书在版编目（CIP）数据

江南园林艺术手绘图志 = Southern Chinese Garden Art Drawings：英汉对照 / 吴肇钊著. —北京：中国建筑工业出版社，2021.4
ISBN 978-7-112-24950-3

Ⅰ. ①江… Ⅱ. ①吴… Ⅲ. ①古典园林－建筑艺术－华东地区－图集 Ⅳ. ①TU-092.2

中国版本图书馆CIP数据核字（2020）第039925号

责任编辑：郑淮兵　王晓迪
书籍设计：锋尚设计
责任校对：王　烨

江南园林艺术手绘图志
Southern Chinese Garden Art Drawings
吴肇钊　著
*
中国建筑工业出版社出版、发行（北京海淀三里河路9号）
各地新华书店、建筑书店经销
北京锋尚制版有限公司制版
北京富诚彩色印刷有限公司印刷
*
开本：787毫米×1092毫米　1/12　印张：28　字数：396千字
2021年4月第一版　　2021年4月第一次印刷
定价：198.00元
ISBN 978-7-112-24950-3
（35708）

吴肇钊 教授

吴肇钊教授，1944年出生，1966年毕业于北京林学院园林系，但在“文革”期间赴中央美术学院进修油画两年。毕业后至今，一直从事园林规划设计与园林理论课题研究，设计作品涉及十二个省、三个直辖市、两个自治区；曾代表国家参加每十年举办一次的国际园林博览会。设计作品“清音园”荣获在德国举办的1993年国际园林博览会“大金奖”。在华盛顿、加拿大、新加坡及我国香港均有设计作品。出版专著《夺天工》《吴肇钊景园建筑画集》《中国园林立意·创作·表现》《瘦西湖园林群历史、艺术与营造》，可贵的是，二十余载的研究成果《园冶图释》巨著（三册），由中国建筑工业出版社以中英双语本隆重出版。任中外园林建设有限公司资深总工程师。现任江苏兴业环境集团风景园林规划设计院总顾问。

1993年德国政府授予荣誉奖章

《夺天工》专著

《园冶图释》专著

《瘦西湖园林群历史、艺术与营造》专著

德国1993年国际园林博览会大金奖

《吴肇钊景园建筑画集》专著

《中国园林立意·创作·表现》专著

Wu Zhaozhao

Professor Wu was born in 1944. In 1966 he graduated from Beijing Forestry College's Landscape Architecture Department. Amidst the "Cultural Revolution", for two years, he engaged in advanced studies of oil painting at Central Academy of Fine Arts. After graduation, he has been engaging in landscape architecture planning and design and theoretical studies in landscape architecture, and his works cover a geographical range over twelve provinces, three municipalities directly under the central government, two autonomous regions. He once represented China to participate an international garden expo that is held once every decade. His garden design work "Qingyin Garden" won grand prize in International Garden Show in Germany in 1993. His design work can be found in Washington DC, Canada, Singapore, and Hong Kong of China. His published books include *Excelling Nature (or Duo Tian Gong), Wu Zhaozhao Garden Architecture Art Album, The Conception, Creation, and Representation of Chinese Gardens, The History, Art, and Construction of the Slender West Lake Garden Clusters*. What's more valuable is the grand publication of the Chinese-English edition of the three-volume magnum opus *Yuan Ye Illustrated,* reflecting his over twenty years research work, published by China Architecture & Building Press. Mr. Wu serves as the senior chief engineer at Landscape Architecture Corporation of China. Now he serves as the general consultant at Jiangsu Xingye Environmental Group Co.,Ltd. Landscape Architecture Planning & Design Institute.

Honorable Award by
German Government, 1993

Excelling Nature

Yuan Ye Illustrated

The History, Art, and Construction of the Slender West Lake Garden Clusters

International Garden Expo
Grand Prize, Germany, 1993

Wu Zhaozhao Garden Architecture Art Album

The Conception, Creation, and Representation of Chinese Gardens

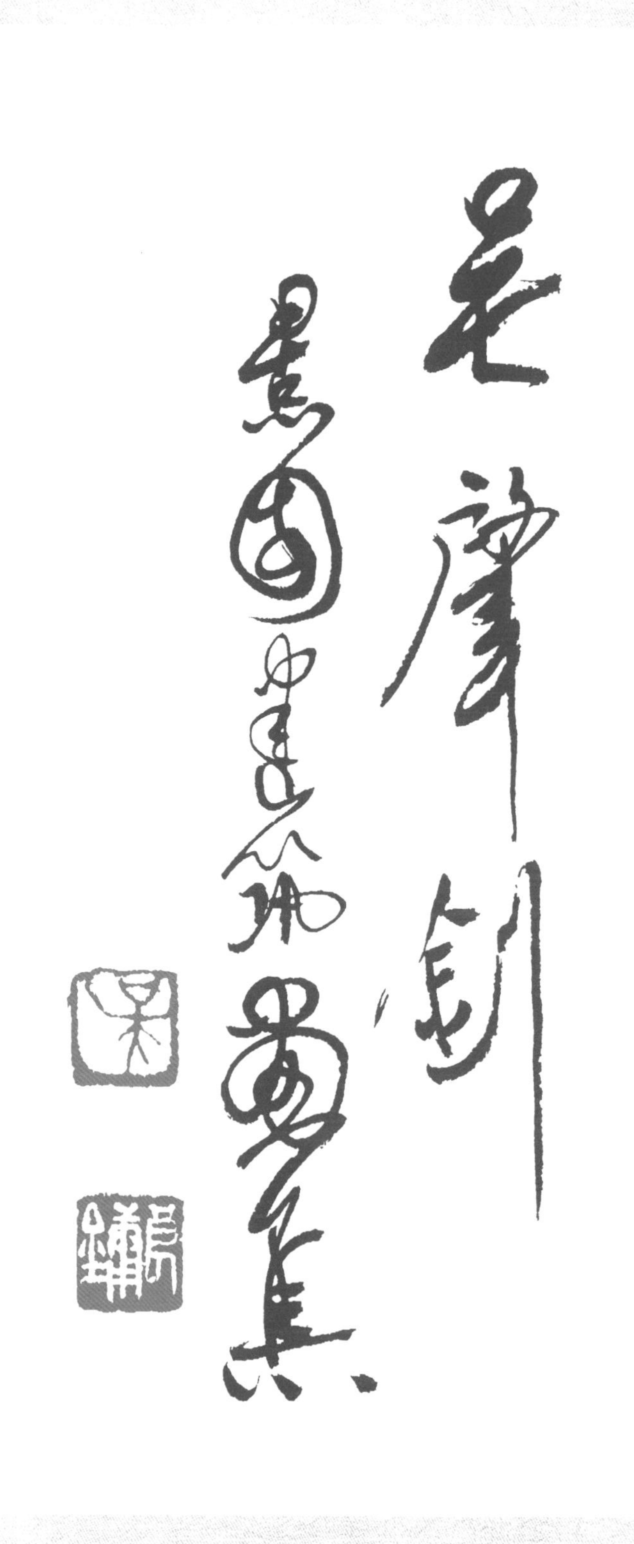

中国科学院 中国工程院
资深院士 **吴良镛**

Calligraphy of the book title by Wu Liangyong, senior dual member of Chinese Academy of Sciences and Chinese Academy of Engineering

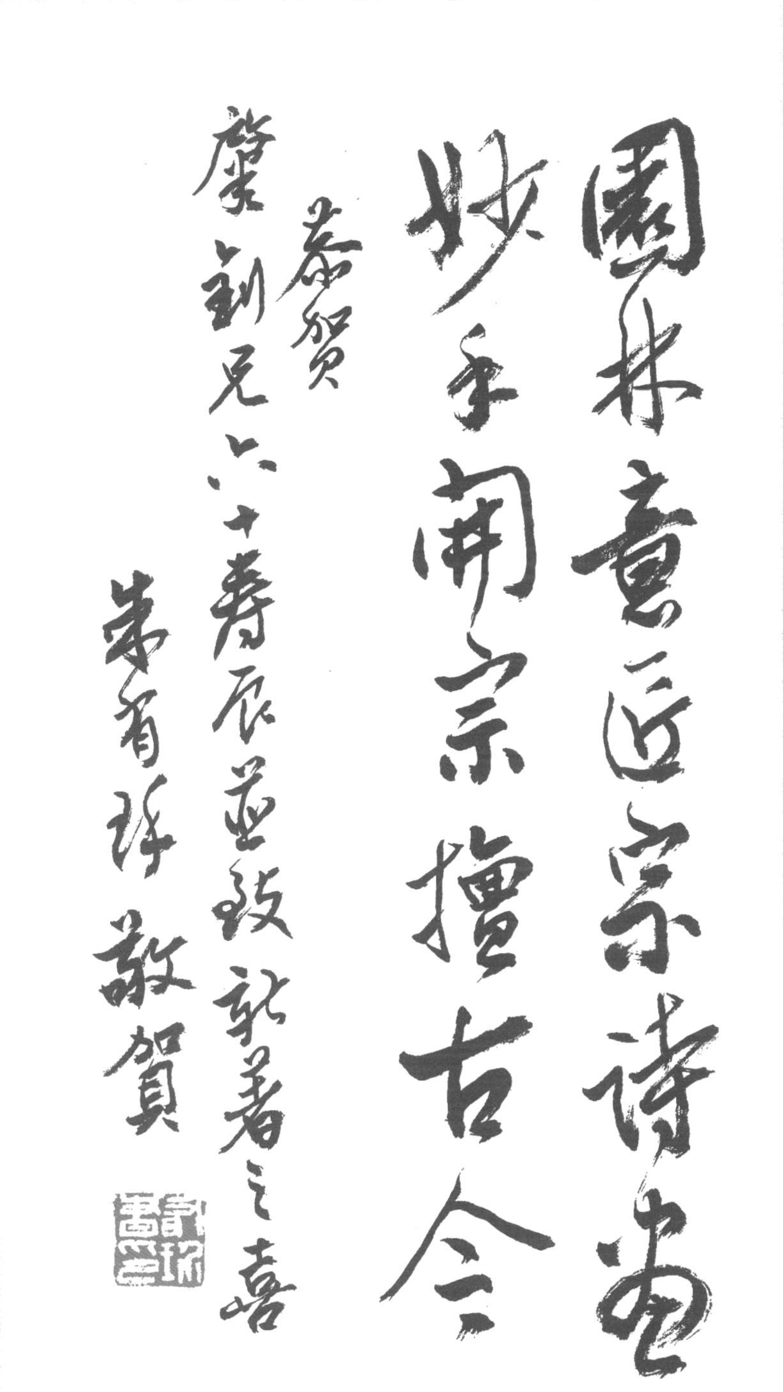

園林意匠宗詩畫
妙手開宗擅古今

恭賀
肇釗兄六十壽辰並致新著之喜
朱有玠敬賀

江苏省首席园林设计大师
朱有玠

Calligraphy of a poem by Zhu Youjie, grand master garden designer, Jiangsu Province

計祖衍傳
優門生因
借體宜妙
相稱文人
山水積盛
譽應時丹
青奪先聲

吳肇釗賢契江南園林
藝術圖誌付梓紀念
時在庚子春日

孟兆禎

中国工程院院士
中国园林学会名誉理事长
孟兆祯

Calligraphy of a poem by Meng Zhaozhen, member of Chinese Academy of Engineering and Honorary Director General of Chinese Society of Landscape Architects

自序

我出生于1944年，1966年毕业于北京林学院园林专业。回想大学期间，恰逢“文革”，鉴于多方原因只能充当“逍遥派”，为不遭批斗，经老师介绍，鼓足勇气去中央美术学院绘毛主席油画像。尽管亦学二年素描、水彩，与美院相比乃天壤之别，只得夜以继日奋战，油画颜料可谓成箩成筐用完。“功夫不负有心人”，经过近两年努力，油画作品也“挤入”1966年的全国美展。学校礼堂及饭厅外墙均挂上我画的巨幅毛主席及中央领导画像，可谓利用二年派性斗争时间多学了一个专业，绘画功底为后来园林设计助了一臂之力。

大学毕业后十分有幸，学术上能得益于几代园林与建筑宗师的培育。早在20世纪70年代，园林界泰斗汪菊渊先生将我选入其主编的《中国古代园林史》编委（我是其中最年轻的），为期五年编撰“江南园林”章节，给我奠定了古典园林理论和实例研究的扎实基础。为此，80年代初，江苏省建委调我赴南京参加编写《江苏园林名胜》一书，恰童寯、陈从周老先生均为顾问，在一年多的时间里，童老《江南园林志》学问的渊博与他惊人的记忆力令人难以忘怀，其创作图稿的精炼与色彩至今仍令学生震撼，并得其引荐求教杨廷宝泰斗。杨老十分耐心地讲图并亲笔示范，其准确工整的画风一直是学子的样板。陈从周老先生文采卓绝为世人公认，他对园林的鉴赏评论亦是难以比拟的。我主笔的石涛“片石山房”修复设计得其指导，受益匪浅。原拟山房长廊内碑刻选用石涛画作，陈老教导“片石山房”已是石涛画再现，碑刻则应为诗书作品，可谓绝妙至极！陈老亦亲自撰写碑记与园名。园林大师朱有玠先生对我的园林设计起到了定位作用，他首次看到学生建成的园子，就赞誉颇有画意，以后每见到学生均告诫“画本再现的风范”是造园成功的准则，致使我至今仍坚持先绘出画本后，再进行施工图设计，以确保画本再现。恩师孙筱祥先生、孟兆祯先生不仅学术渊博，而且多才多艺，让学生吸收了各具特色的丰富营养，他们的厚爱加速了学生的成才。孟先生一直是学生园林学科的“辞源”，本书掇山章节先生均页页过目，亲笔修正，实令学生感激万分。

正当我年富力强之际，甘伟林司长、王泽民总经理将我调入中国对外园林建设总公司任总工，得以参与海外园林工程设计并主持施工，德国、美国、法国、加拿大、日本、新加坡、马耳他等国家以及我国香港均留下了我的园林作品与设计。在完成中国园林在海外传播任务的同时，自觉地更翔实考证、研究祖国古典园林技艺，并坚持以中国画的手法绘园林效果图，否则难以应答各方洋学者提出的“尖攒”、溯源等学术难题，以及现场画图和举办个人画展的要求。现在看来，这些对个人学术的成长饱含积极意义。

学生研究江南古典园林的功底，更得益于四十余载设计建筑古典园林的实践，诸如石涛“片石山房”的复建，扬州瘦西湖“卷石洞天”“白塔晴云”“二十四桥景区”“静香书屋”等古园林的复建等，被孟兆祯院士誉为有的作品“已达到新中国成立后全国一流的先进水平”。1989年江苏省园林优秀作品评选时，曾囊括前三名。此后，宗师们的鼓励，同行、同仁频频敦促学生完成《园冶图释》科研成果，出版社亦寄予厚望并决定以中英文对照出版。哪知道该书出版后，日本亦向我国提出翻译日文版的申请，今年我亦在申请书上签上了“同意”。

由于《园冶图释》的出版，众多园林部门以及大学请我做专题学术报告，并提出要求，希望我能钢笔手绘江南园林精华景点，这样既可以读懂江南古典园林，亦可以起到“画帖”的作用。众望所托，我只得放下75岁高龄的包袱，高价配置高档的眼镜，绘制《江南园林艺术手绘图志》，既是向传道授业宗师致谢，向之汇报；同时亦是让更多中外人士读懂江南古典园林，共同“借古开今”！鉴于是探索的开篇，不足之处敬请同行不吝斧正。

本书的出版得到江苏兴业环境集团董事长的大力支持，在此表示感谢！

吴肇钊

2019年9月1日

PREFACE /

I was born in 1944 and graduated in 1966 from Beijing Forestry College in Landscape Architecture major. I recall that in my college time, in the middle of "the Cultural Revolution", due to many reasons I could only resort to acting as a "happy camper" who took no sides in order to avoid being harshly criticized with physical abuse. Through recommendation of my professor I summoned enough courage to take classes at Central Academy of Fine Arts to learn canvas portrait oil painting of Chairman Mao. Although I had already spent two years learning sketch and watercolor painting, comparing with the works done by their own college majors, I could still see significant difference. I had to spend days and nights to improve my skills, consuming buckets after buckets of oil paint. Hard work paid off. After nearly two years of laborious work, my artwork finally entered national art exhibitions. My canvas paintings of Chairman Mao and other national leaders were hanged at my college auditorium and walls outside campus canteens. So during these two years of tumultuous factional conflicts I wasted no time and mastered a new major, and that painting skill helped me tremendously in my future landscape garden design work.

After graduating from college, academically and professionally I was very fortunate to get nurtured under generations of great masters in garden design and architecture. As early as the 1970s, Wang Juyuan, a leading authority in Chinese landscape architecture, invited me to join the editorial board of *Chinese Ancient Garden History,* of which he was the editor-in-chief (and I was the youngest board member). Five years of working on the project on Southern Chinese Garden chapter had laid a solid foundation for me in Chinese classical garden theory and practice. Because of this project, in the early 1980s, Jiangsu Construction Board appointed me to travel to Nanjing and work on compiling *Jiangsu Garden Attractions*. During that period of time, we had senior consultants Tong Jun and Chen Congzhou. During more than a year's time working with them, I was impressed by Mr. Tong's incisive scholarship on *Annals on Southern Chinese Gardens*; his concise and colorful design work still remain awe-inspiring and amaze me to this day. I was introduced by him to learn from Yang Tingbao, a leading authority on architecture. Mr. Yang was very patient in design work interpretation and he showed design examples in person by himself. His accurate and neat style has been a model for me to follow. Mr. Chen Congzhou's literary talent is well-known and unsurpassed, and his appreciation and critiques on garden art are marvelous. The restoration design of Shi Tao's "Pianshi Shanfang" led by the author had been guided by Mr. Chen and I had benefited from him greatly. Originally, Shi Tao's paintings were supposed to be the contents of the stele inscriptions inside the long corridor; however, Mr. Chen pointed out that "Pianshi Shanfang" itself was already the realization of Shi Tao's paintings and as such, the stele inscriptions should show his poetry and other writings. What a brilliant idea! Mr. Chen then wrote the record of the stele inscription and named the garden. The garden master Zhu Youjie played a positioning role on the author's landscape garden design career. When he first saw a garden that I built, he complimented that it was pretty picturesque. Later whenever he saw me he had always admonished that the success of a garden design is "the realization of an artwork". This made me keep making an artistic drawing before making a constructed one to ensure that the artistic version is realized. My mentors Sun Xiaoxiang and Meng Zhaozhen are not only erudite but also multi-talented. They let their students absorb all the rich nutrients of different characteristics and their deep love made their students qualified for garden design work sooner. Professor Meng is respectively regarded as a live dictionary by landscape architecture students. He personally reviewed every page on the chapter about rock and mountain building and for that I appreciate immensely.

During the prime of my life, Gan Weilin, China's Construction Ministry's director, and general manager Wang Zemin assigned me chief engineer's position in China's Overseas Landscape Architecture Construction Company. This gave me opportunities to participate oversea project designs and constructions and I had left my work in Germany, the United States, France, Canada, Japan, Singapore and Malta and so on. While propagating Chinese garden art overseas, I consciously conducted detailed researches and studies about classic Chinese garden art and insisted on using methods in Chinese landscape painting in my designed garden renderings; otherwise it would be difficult to field various critical academic questions, some of them dating back to ancient times, posed by international scholars. They also requested me to make artistic renderings at the site and host personal art shows. Looking at those events retrospectively, I now realize that they were really beneficial for my personal academic growth.

Armed with the skills to study classical southern Chinese gardens, and more importantly, I had more than 40 years of experience designing and building classical gardens, such as the renovation work with Shi Tao's "Pianshi Shanfang", the Slender West Lake's "Juanshi Dongtian", "Baita Qingyun", "Twenty-four Bridges Scenic Zone", and "Jingxiang Study" in Yangzhou, thus won even Academician Meng Zhaozhen's compliments on some of my work as reaching the "first class status after the founding of the new China". In 1989, I had won all top three prizes in Jiangsu Province's Outstanding Landscape Garden Design. Since then, after encouragement from great garden masters and frequent urgings of my peers and colleagues, I planned to finish my research project on *Yuan Ye Illustrated*. With high hopes, the press decided to publish the book in Chinese and English bilingually. What I did not expect is that there is much feedback and response after this publication. Japan made the request to have a Japanese edition and this year I signed the agreement on their application.

Due to the publication of *Yuan Ye Illustrated*, many landscape architecture departments and universities invited me to give academic lectures on special topics, and they asked if I could make manual pen drawings on selected southern Chinese gardens so that they can be understood easily, and the drawings can also be served as model artwork. Upon people's expectations, I had to put away my burden of being 75, equipped with the highest grade eyeglasses, and started hand-drawing *Southern Chinese Garden Art Drawings*. One purpose is to show my appreciation to my mentors; another is to report my work to them. This is also for the purpose of understanding southern Chinese classical gardens, so that we could collectively get inspired by learning from the past. Considering this is just a start, there must be some mistakes made and I would welcome my readers' comments for improvements.

I would like to express my sincere thanks for the support I received from the chairman of the board of Jiangsu Xingye Environment Group.

Wu Zhaozhao

September 1, 2019

目录

总图系列

景点系列

理石系列

海外系列

TABLE OF CONTENTS

A / GENERAL SERIES

B / SCENIC SPOT SERIES

C / ROCKERY ARRANGEMENT SERIES

D / OVERSEA SERIES

总图系列

A

GENERAL SERIES

苏州拙政园鸟瞰图 | Bird's Eye View of the Humble Administrator's Garden in Suzhou

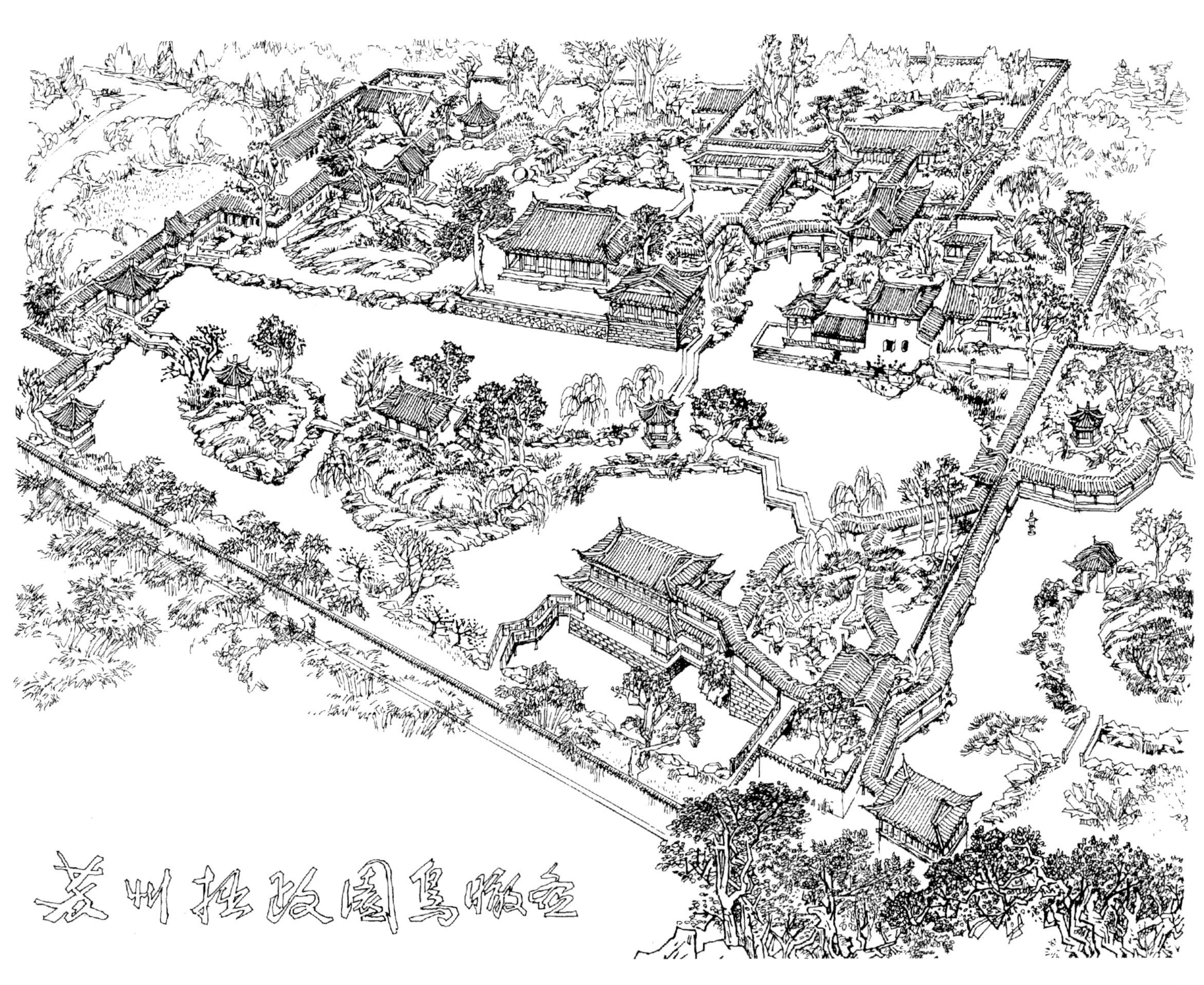

苏州拙政园中部鸟瞰图（一）

Mid-Section Bird's Eye View of the Humble Administrator's Garden in Suzhou (I)

苏州拙政园中部鸟瞰图（二）

Mid-Section Bird's Eye View of the Humble Administrator's Garden in Suzhou (II)

苏州拙政园局部鸟瞰图

Bird's Eye View of Part of the Humble Administrator's Garden in Suzhou

苏州拙政园小沧浪水院鸟瞰图

Bird's Eye View of Small Canglang Water Court at the Humble Administrator's Garden in Suzhou

苏州怡园鸟瞰图（一）

Bird's Eye View of the Yi Garden in Suzhou (I)

苏州怡园鸟瞰图（二） Bird's Eye View of the Yi Garden in Suzhou (II)

苏州沧浪亭鸟瞰图

Bird's Eye View of Canglang Pavilion in Suzhou

苏州网师园（古代界画法）

The Net Master's Garden in Suzhou (Ancient Ruler Drawing)

苏州网师园中部
鸟瞰图

Mid-Section Bird's Eye View of the
Net Master's Garden in Suzhou

苏州留园石林小院剖视

Section View of the Rock Forest Courtyard of the Lingering Garden in Suzhou

苏州留园冠云峰庭院鸟瞰图

Bird's Eye View of Guanyun Rock Courtyard at the Lingering Garden in Suzhou

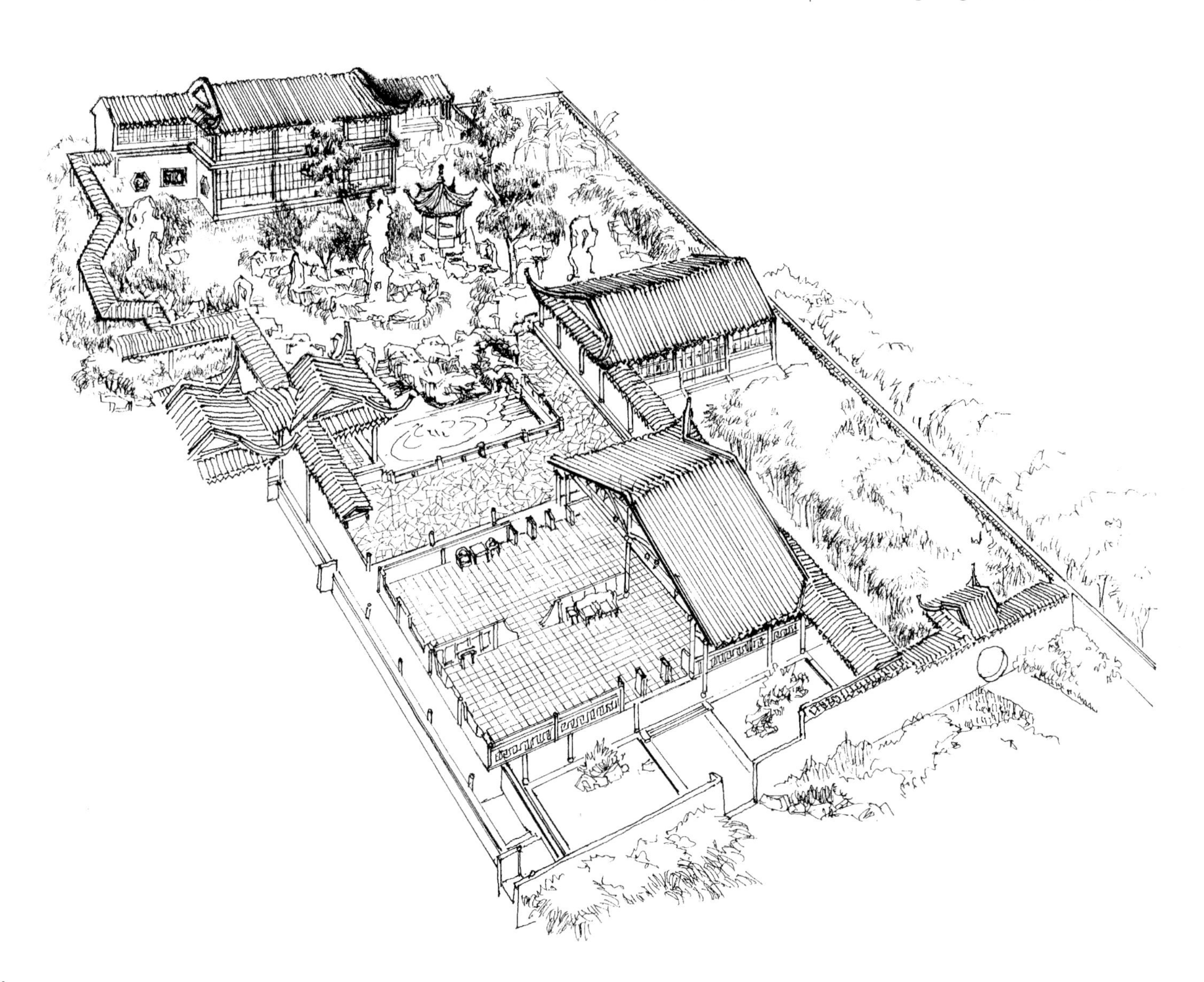

苏州狮子林鸟瞰图 | Bird's Eye View of the Lion Grove Garden in Suzhou

苏州畅园
以水池为中心，并环绕水池四周布置建筑，从而具有向心与内聚的感觉。

The Chang Garden in Suzhou
Centered around the water body, the buildings are set around, to forge a central pulling force.

苏州退思园鸟瞰图

Bird's Eye View
of the Retreat &
Reflection Garden
in Suzhou

苏州拥翠山庄鸟瞰图

Bird's Eye View of Yongcui Villa in Suzhou

苏州王洗马巷古宅书房剖视图

Section View of Ancient Study at Wangxima Lane in Suzhou

苏州壶园鸟瞰图 | Bird's Eye View of the Kettle Garden in Suzhou

苏州壶园剖视图 | Section View of the Kettle Garden in Suzhou

苏州天平山高义园局部庭院 | Partial Courtyard of the Gaoyi Garden at Mount Tianping in Suzhou

无锡惠山云起楼庭院 | Courtyard of Yunqi Building at Mount Hui in Wuxi

无锡寄畅园秉礼堂与庭院鸟瞰图 | Bird's Eye View of Bingli Hall and Courtyard at Jichang Garden in Wuxi

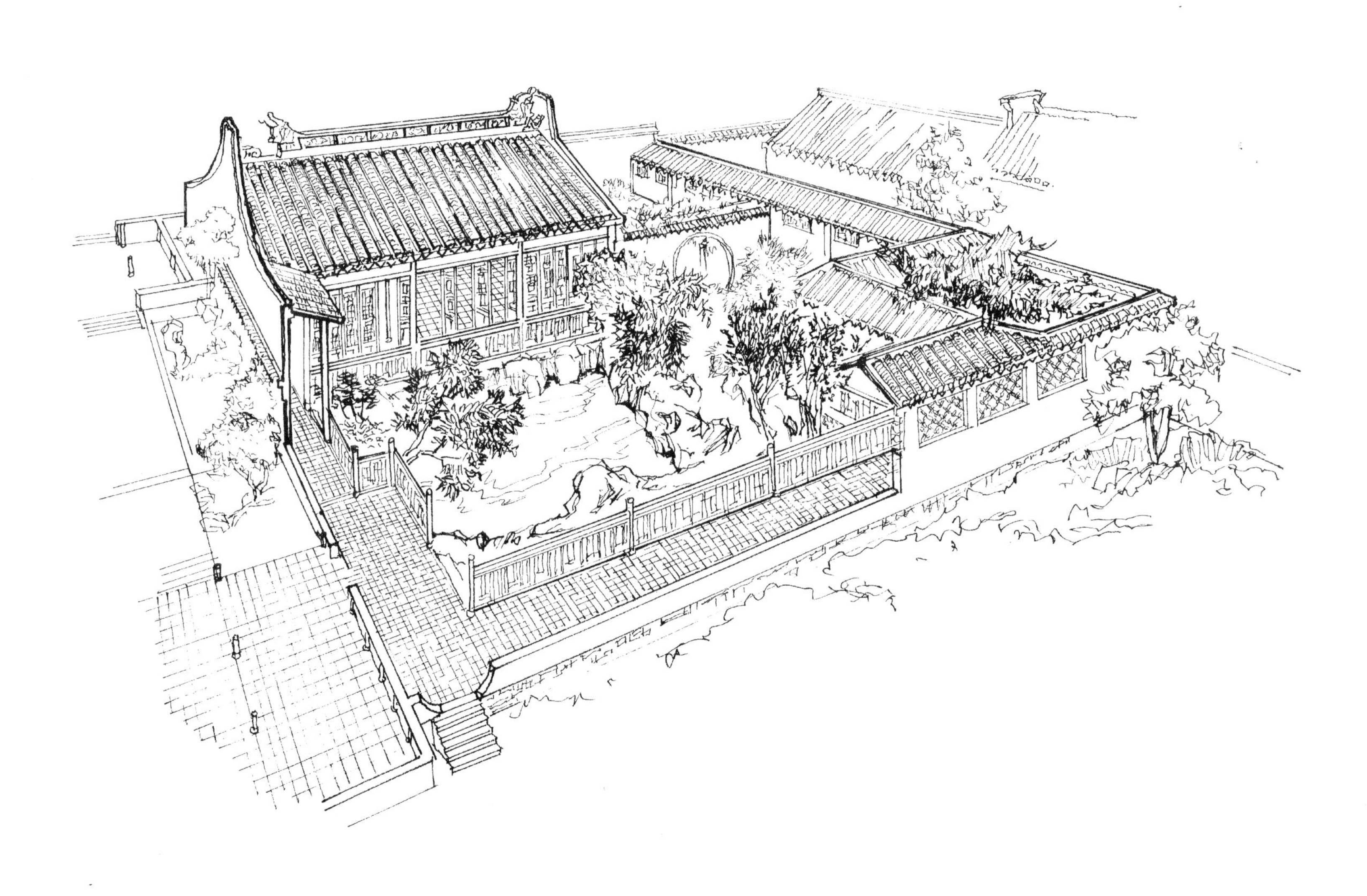

杭州西湖小瀛洲、三潭印月鸟瞰图

Bird's Eye View of Miniature Yingzhou and Three Pools Mirroring the Moon at West Lake in Hangzhou

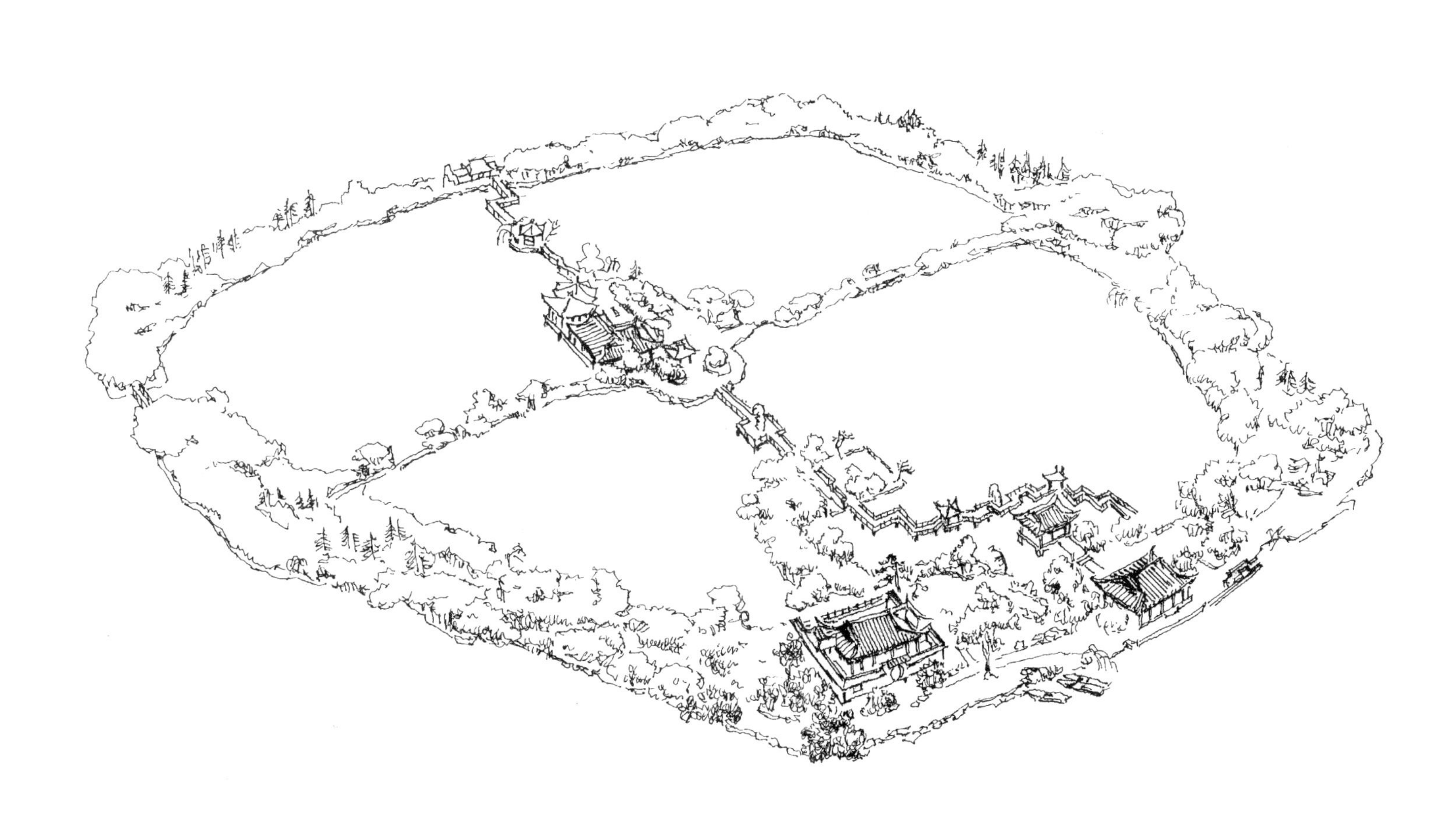

杭州文澜阁庭院鸟瞰图

Bird's Eye View of Wenlan Pavilion Courtyard in Hangzhou

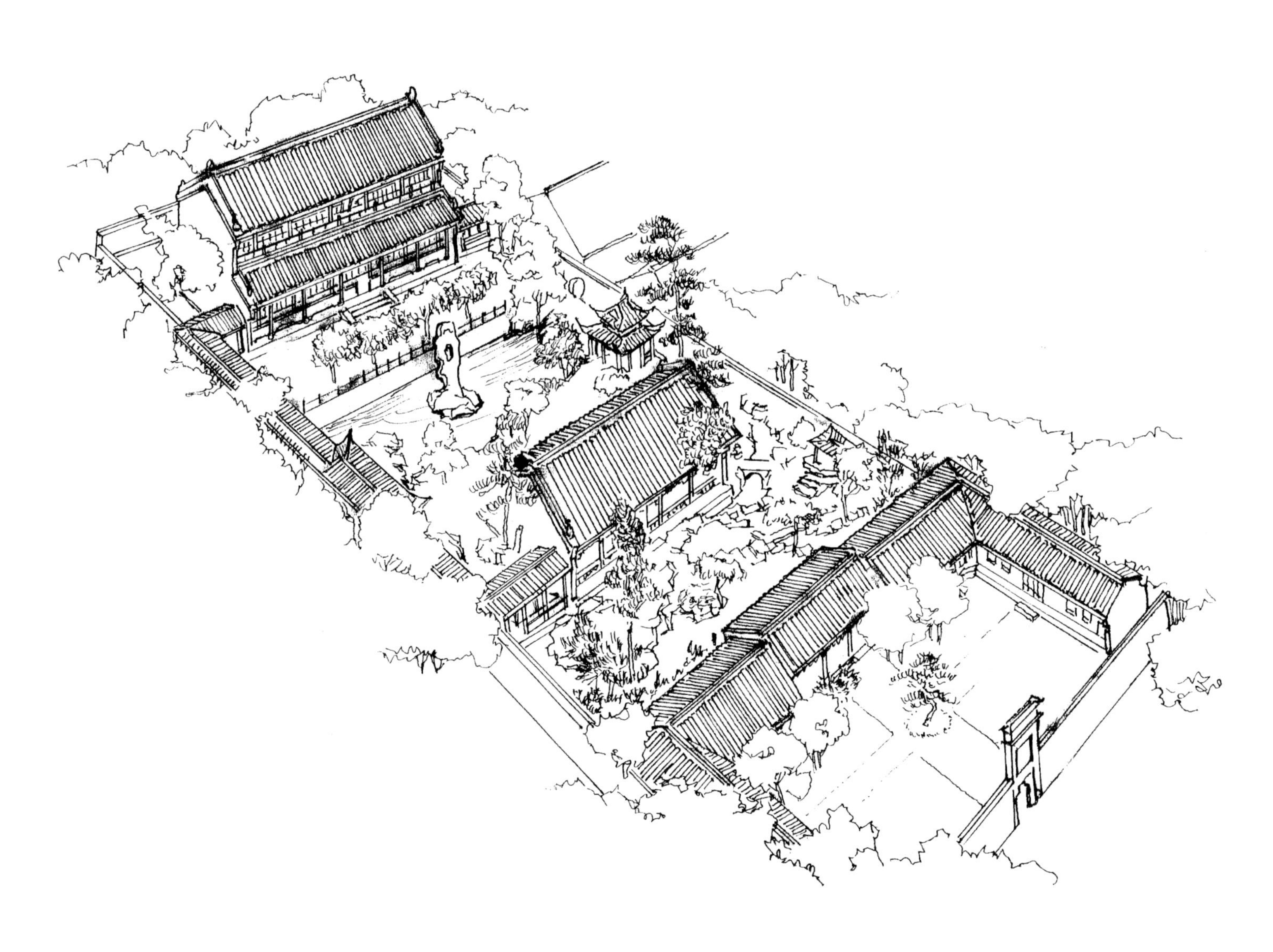

扬州瘦西湖二十四桥景区鸟瞰图（著者设计）

Bird's Eye View of Twenty-four Bridges Scenic Areas at Slender West Lake in Yangzhou (Designed by the Author)

嘉兴烟雨楼鸟瞰图 | Bird's Eye View of Misty Rain Building in Jiaxing

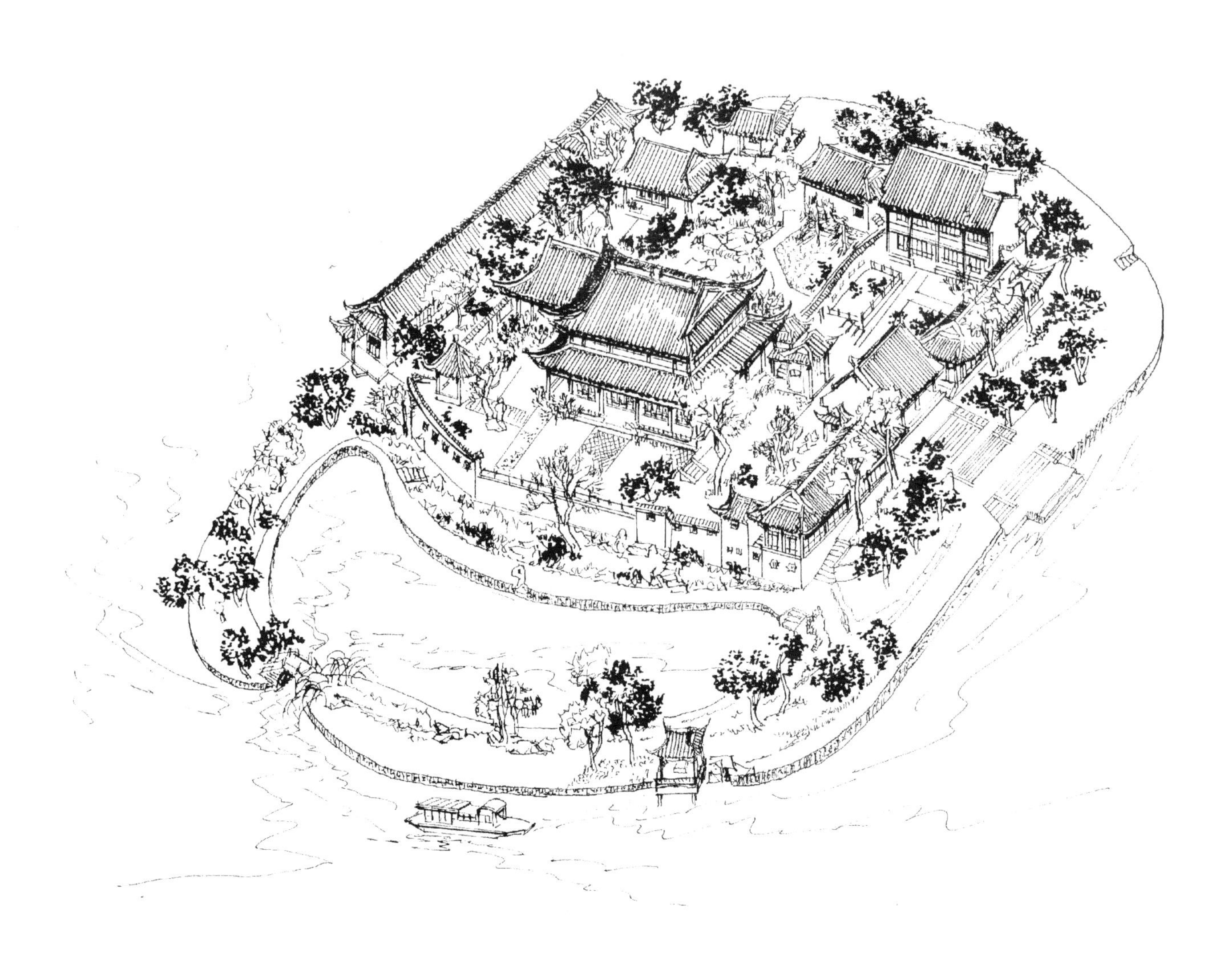

绍兴兰亭鸟瞰图 | Bird's Eye View of Lan Pavilion in Shaoxing

淮安恢台园（著者复原设计）
为探花夏曰瑚始建，面城带水，水阔处可百丈。

Image of Huitai Garden in Huai'an (Restoration Designed by the Author)
Initially built by Xia Yuehu who got the third place in the highest imperial examination. Facing the city and with a water body, with the widest at a hundred yards.

淮安柳衣园（著者复原设计）
为进士吴准及其父亲吴进建造。

Liuyi Garden in Huai'an (Restoration Designed by the Author)
Built by Wu Zhun, a successful candidate in the highest imperial examination, along with his father Wu Jin.

淮安带柳园（著者复原设计）
为贡生程茂建，园不大而设计曲折。

Dailiu Garden in Huai'an (Restoration Designed by the Author)
Built by tribute student Cheng Mao. The garden is not large but its structure is convoluted.

淮安荻庄（著者复原设计）
园主程沆为进士，此园作为乾隆南巡御花园。

Di Villa in Huai'an (Restoration Designed by the Author)
The owner Cheng Hang was a successful candidate in the highest imperial examination. This village served as Emperor Qianlong's imperial garden during his southern China tour.

淮安曲江园（著者复原设计）
位于萧湖北侧湖面，为进士张新标建造。

Image of Qujiang Garden in Huai'an (Restoration Designed by the Author)
Located in the north of Lake Xiao. Built by Zhang Xinbiao, a successful candidate in the highest imperial examination.

天津红学家纪念馆
（著者设计）

A Dream of Red Mansions Scholar Memorial in Tianjin (Designed by the Author)

临沂盆景博览园
（著者设计）

Bonsai Expo Park in Linyi
(Designed by the Author)

深圳弘法寺增建方案（泉水系列）（著者设计）
寺庙住持本焕长老对方案十分满意，时逢本焕长老恰100岁，故著者请他在总图上签名，今用此图意在希望著者与读者均达到高寿福德之境界。

Extension Design of Hongfa Temple in Shenzhen (Spring Series) (Designed by the Author)
Elder abbot Ben Huan of the temple was very satisfied with the design. Considering that the abbot was exactly 100 years old, the author asked him to sign on the drawing. Now the author would like to use this drawing to let both the author and the readers enjoy the state of longevity and good fortune.

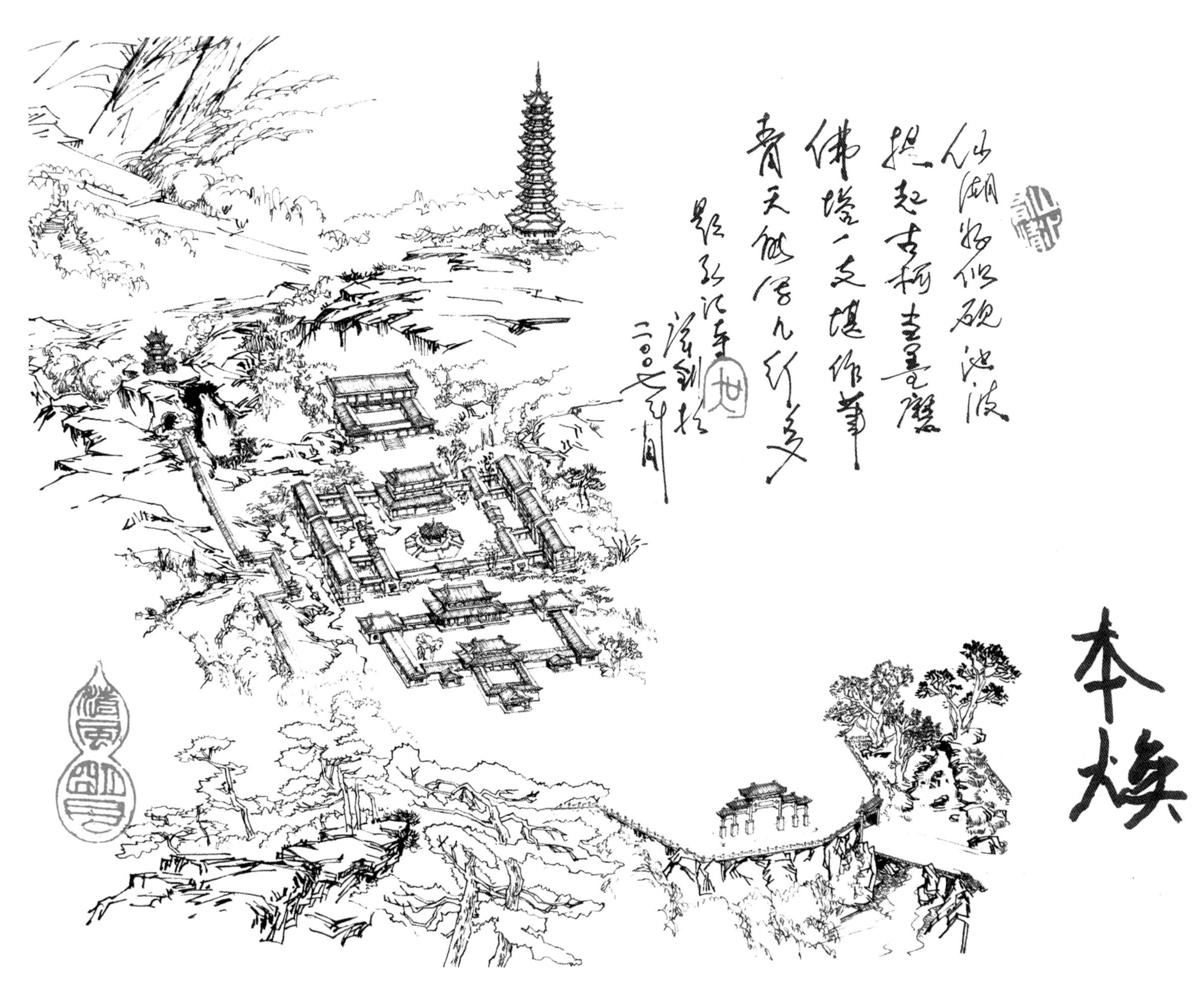

古建综合博览园（著者设计）| General Expo Park of Ancient Architecture (Designed by the Author)

李清照陈列馆“庭院雪”景区（著者设计）| "Court Snow" Scenic Spot at Li Qingzhao Exhibit Hall (Designed by the Author)

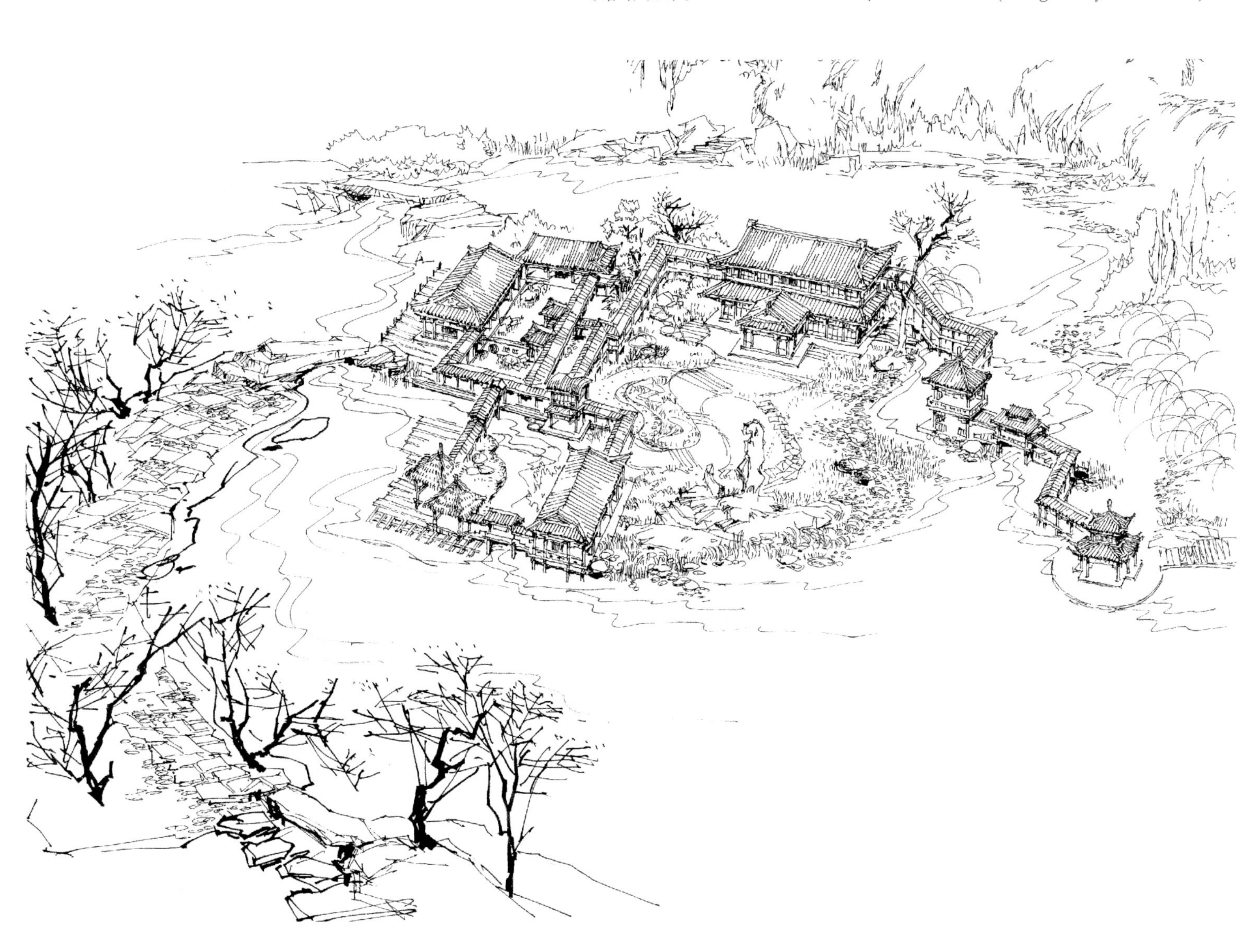

仿绘建筑大师尚廓的“水院清凉图”

Imitation Drawing of "Cool Water Court" by master architect Shang Kuo

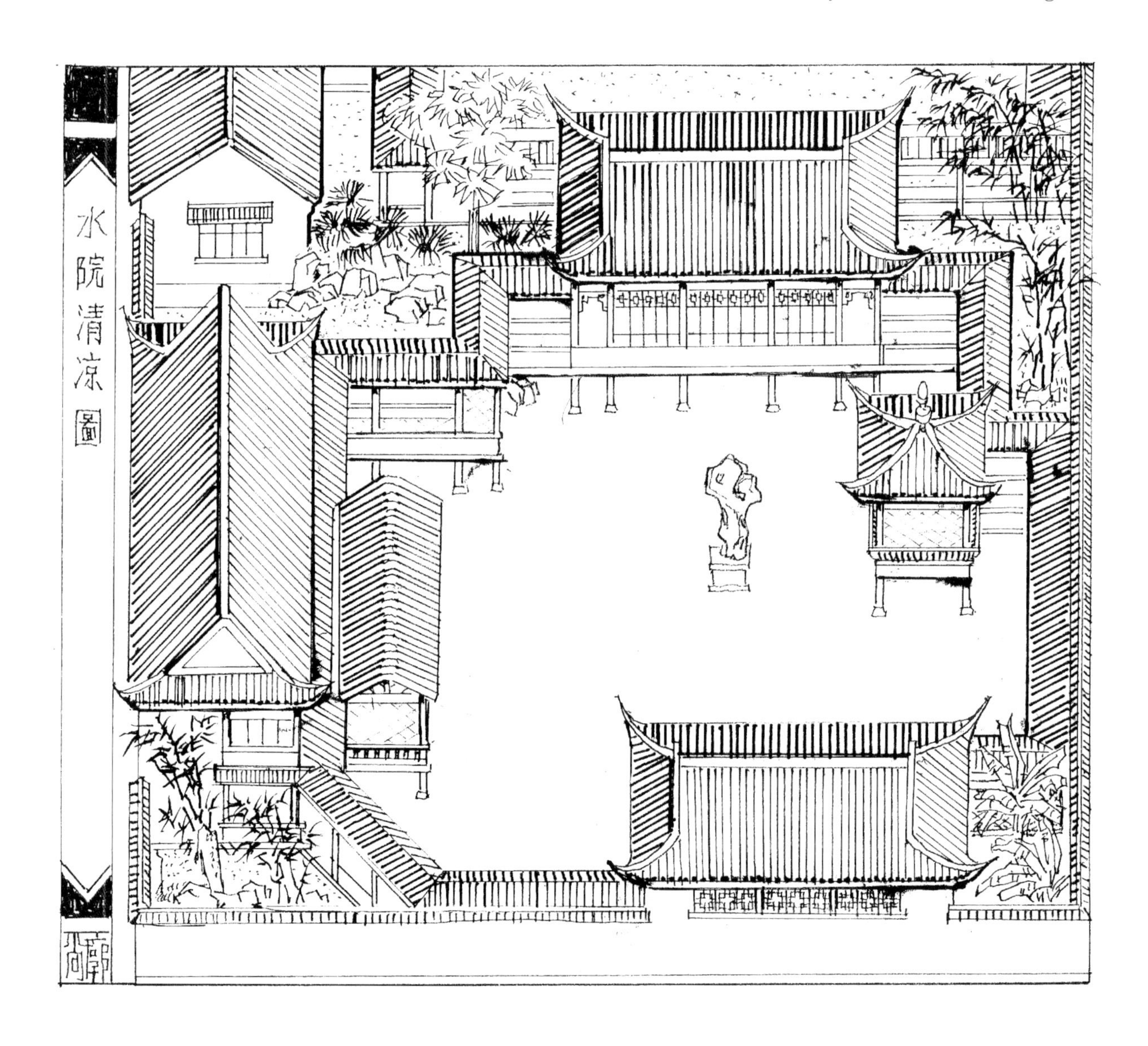

水苑嬉月 | Water Park Frolicking Moon
（著者设计） | (Designed by the Author)

荷蒲熏风
（著者设计）

Breeze over Lotus Pond
(Designed by the Author)

新加坡“中国园”方案设计
（仿绘彭一刚院士效果图）

"Chinese Garden" Design in Singapore (Imitation Drawing after Academician Peng Yigang)

德国清音园

此图系著者代表中国参加1993年在德国举办的国际园林博览会时，中国参展园“清音园”的设计方案鸟瞰图，获博览会董事会较高评价，按此图建成的“清音园”，荣获1993年国际大金奖，并获德国政府颁发的荣誉奖章。

Qingyin Garden in Germany

This drawing is design concept bird's eye view of "Qingyin Garden" drawn by the author to represent China to participate the International Garden Expo held in Germany in 1993. It achieved high acclaim at the event. "Qingyin Garden" built based on this drawing won International Grand Gold prize and also won an honorable prize from the German government.

景点系列

B

SCENIC SPOT SERIES

个園

苏州拙政园远借北寺塔
苏州拙政园远借北寺塔，是借景的佳作。

Beisi Pagoda Scene Distantly Borrowed by the Humble Administrator's Garden in Suzhou.
It is an excellent example of a borrowed scene.

苏州拙政园月洞门

由拙政园枇杷园透过月洞门北望岛山上“雪香云蔚亭”。

The Round Gate inside the Humble Administrator's Garden

From inside the Loquat Garden in the Humble Administrator's Garden, one can peek to the north through the round garden gate towards Xuexiang Yunwei Pavilion on top of island hill.

苏州拙政园透过倒影楼窗口看宜两亭

Yiliang Pavilion Seen through a Window of Reflection Building from the Humble Administrator's Garden

苏州拙政园透过宜两亭窗口看倒影楼

Reflection Building Seen through a Window of Yiliang Pavilion from the Humble Administrator's Garden

苏州拙政园西部水上浮廊、扇面亭及宜两亭
丰富的层次引人入胜。

Floating Corridor, Fan-Shaped Pavilion and Yiliang Pavilion in West Part of the Humble Administrator's Garden
Rich layers of garden scenes are very alluring.

苏州拙政园小飞虹横卧水上
其南构成一个独立的水院。

The Small Flying Rainbow Bridge Flying over Water at the Humble Administrator's Garden
Its south part is made to form an independent water courtyard.

苏州拙政园小飞虹廊桥
分隔水面，丰富层次。

The Small Flying Rainbow Bridge at the Humble Administrator's Garden in Suzhou Dividing the water body, it creates rich layers of scenes.

苏州拙政园枇杷园

作为园中园采用“云墙”围合，分隔空间，本身起伏转折，犹如蛇行之蜿蜒，是一处优美的景观。

Loquat Garden at the Humble Administrator's Garden in Suzhou

As the garden within a garden separated by "cloud walls", it serves the purpose of dividing spaces and creating rhythm as if serpent winding, as well as beautiful landscape scenes.

苏州拙政园枇杷园以云墙分隔成的小院
院内亭周边种枇杷树，是一个组合得非常自由和巧妙的园中之园。

Small Courtyard Separated by the Cloud Walls of the Loquat Garden at the Humble Administrator's Garden in Suzhou
Loquat trees are planted around the pavilion. This is a clever design for a garden within a garden with free style arrangement.

苏州拙政园扇面亭 | Fan-shaped Pavilion at the Humble Administrator's Garden in Suzhou

苏州拙政园扇面亭西南面门洞、背面及东北面门洞

扇面亭西南门洞看三十六鸳鸯馆的对景效果。
扇面亭背面的扇形窗口可窥见园西北的浮翠阁。
通过东北面门洞看侧影楼的对景效果。

Southwest Door Opening, the Back Side and the Northeast Door Opening at the Fan-Shaped Pavilion at the Humble Administrator's Garden in Suzhou

The opposite scenes of the Thirty-six Mandarin Duck Building through the door.

At the back of the fan-shaped window, one can peek through the window to see Fucui Pavilion in northwest part of the garden.

Through the door opening in the northeast, one can get the opposite scene of Ceying Building.

苏州拙政园扇面亭、倒影楼与周围景物构成的画面

Mini Picture Park Composed by Fan-Shaped Pavilion, Reflection Building, and the Surrounding Scenes at the Humble Administrator's Garden in Suzhou

苏州拙政园听雨轩
为玲珑馆东水池畔营黄石山冈、种植芭蕉、寄情“雨打芭蕉”的曲韵。

Veranda for Listening to the Rain at the Humble Administrator's Garden in Suzhou
It is located in the east of Linglong Pavilion at the pond side, which is constructed with yellow rocks and planted with Chinese bananas, corresponding to the famous melody "raindrops hitting banana leaves".

苏州拙政园听雨轩庭院剖面图 | Section View of Listening to the Rain Pavilion Courtyard at the Humble Administrator's Garden

苏州拙政园中部小院
小院相套，水系萦回，层次丰富。

Small Courtyard in the Middle Section of the Humble Administrator's Garden in Suzhou
Yard in a yard with intricate water channels, which makes rich layers of garden scenes.

苏州拙政园香洲北立面图 | North Facade of Xiangzhou at the Humble Administrator's Garden in Suzhou

苏州拙政园海棠春坞鸟瞰 | Bird's Eye View of Haitang Chunwu at the Humble Administrator's Garden in Suzhou

苏州网师园“月到风来亭”

凌水驾波的“月到风来亭”是园中美景之一，最宜月夜观赏皓月当空，微风轻拂，水中月影浮动，令人流连忘返。

Yuedao Fenglai Pavilion at the Net Master's Garden in Suzhou

Over water, Yuedao Fenglai Pavilion is one of the fine scenes of the garden. Enjoy it most at a moon-lit night, with breeze blowing and moon reflecting in water, making you indulging with pleasures and reluctant to leave.

苏州网师园撷秀楼北庭院竹石小品 | Bamboo and Rock Arrangements at Xiexiu Building's North Court at the Net Master's Garden in Suzhou

苏州网师园竹外一支轩及射鸭廊
使室内外空间丰富多变，是休憩佳处。

Zhuwai Yizhi Pavilion and Duck Shooting Corridor at the Net Master's Garden in Suzhou
The pavilion and the corridor make rich interior-exterior spatial variations, a great place to relax and rest.

苏州网师园竹外一枝轩西主立面
高低错落，景色诱人。

West Main Facade of Zhuwai Yizhi Pavilion at the Net Master's Garden in Suzhou
Its west main facade is consist of components at varied heights, creating an attractive scene.

苏州留园明瑟楼下可以观赏东北部园景 | Northeast Scenes Seen from Mingse Building at the Lingering Garden in Suzhou

苏州留园鸳鸯厅

以北园林是以冠云峰为主题来布局，画意盎然。

Mandarin Duck Hall at the Lingering Garden in Suzhou

The main subject of the garden north of the building is Guanyun Rock, and the whole garden is laid out based on this subject, very picturesque.

苏州留园濠濮亭
建于水上，以庄子“濠梁观鱼、濮水垂钓”典故而名。

Haopu Pavilion inside the Lingering Garden in Suzhou
It was built above water, which was named after the allusion "watching fish at Haoliang, and fishing at Pushui" by Zhuangzi.

苏州留园自远翠阁
看中部曲廊、云墙，
可谓蜿蜒曲折。

From Yuancui Pavilion, at the Lingering Garden in Suzhou
One looks at the winding corridor and the cloud garden wall, with twists and turns.

苏州留园林泉耆硕之馆内部透视（由西望东） | Interior Perspective Rendering of Linquan Qishuo Pavilion at the Lingering Garden in Suzhou (Looking from the West towards the East)

苏州留园五峰仙馆与庭院 | Wufengxian Pavilion and Courtyard Rendering of the Lingering Garden in Suzhou

苏州留园“静中观”东视石林小院画面清秀优美。

"Quiet View" Rock Woods Courtyard at the Lingering Garden in Suzhou Looking east towards the rock grove court yard One is facing a clear and melodious picture.

苏州留园“古木交柯” | Gumu Jiaoke at the Lingering Garden in Suzhou

苏州留园冠云峰庭院北视剖面全景 | Section North View of Guanyun Rock at the Lingering Garden in Suzhou

苏州狮子林花篮厅西的真趣亭
专为悬挂乾隆皇帝手书“真趣”金匾而建，于此观赏园中山水，可得自然真趣。

Genuine Interest Pavilion at West of Hualan Hall at the Lion Grove Garden in Suzhou
This was specifically built for hanging Emperor Qianlong's handwriting "Genuine Interest". When you are enjoying the landscape scene here, you would get the genuine interest from nature.

苏州狮子林古五松园
庭院湖石峰小品
互为呼应，相得益彰。

Courtyard Lake Rockery Peak in Ancient Five Pine Garden at the Lion Grove Garden
The rockery and the yard correspond each other well and can bring the best in each other.

苏州狮子林古五松园向东对景
为赏石佳处。

The East Facing View Point of Ancient Five Pine Garden at the Lion Grove Garden
It is the best place to enjoy the rockery.

苏州狮子林问梅阁北

瀑布自山巅飞泻而下，直往洞底潺然有声，有做假成真的效果。

North of Wenmei Pavilion at Lion Grove Garden in Suzhou

The waterfall plummets from the rockery peak directly into the pit bottom, making a resounding sound that mimics the wild nature.

苏州狮子林屹立在池畔假山上的修竹阁 | Slender Bamboo Pavilion Standing at Pond Side Rockery at Lion Grove Garden in Suzhou

苏州艺圃浴鸥庭院
以粉墙构成空间，
各类植物色彩鲜明，
小池旁湖石点缀，
幽雅静谧。

Bathing Gulls Courtyard at the Garden of Cultivation in Suzhou
With white paint garden wall forming space, various plants with vibrant colors and lake rocks dotting the pond, it feels quiet and elegant.

苏州艺圃水池西岸湖石池山
山水相依，花木掩映。

Lake Rockery Pond and Hillock at the West Shore of the Pond at the Garden of Cultivation in Suzhou
Flower and plant shaded rockeries and water are well integrated.

苏州环秀山庄
仅一亩之地，岩峦耸翠，池水映天，气势磅礴，巧夺天工。“咫尺山林”如真山，入画意，堪称“艺术精品”。

The Mountain Villa with Embracing Beauty in Suzhou
Its size is only one mu, yet it embodies a wonderful article excelling nature by creating a majestic rocky landscape with reflecting waters. Miniature in scale, it looks real and picturesque, and can be categorized into the elite class of fine art.

苏州环秀山庄峦、坡、阶、麓、岩综合景观 | A Combination of Mountain Range, Slopes, Mountain Foot, and Rocky Landscape at the Mountain Villa with Embracing Beauty in Suzhou

苏州环秀山庄石壁
占地甚微，却有洞、壑、涧、谷、悬崖，玲珑有致，自成一体，又融入主山之中。

Rock Wall at the Mountain Villa with Embracing Beauty in Suzhou
Although it only occupies a tiny piece of land, it contains caves, gullies, ravines, canyons, and cliffs; it's exquisite, self-contained, and well-integrated into the main vein of the mountain.

苏州环秀山庄涧水与廊桥
景致高远深邃。

Gully Water and the Corridor Bridge at the Mountain Villa with Embracing Beauty in Suzhou
It is with lofty and profound feeling.

苏州耦园引城河水入园

利用高差堆叠黄石山成涧谷造景，架折桥串联两壁，远处听橹楼屹立城河边借景，是巧妙将园融入环境的佳作，亦是江南园林最杰出的实例。

Urban Channeling River at the Couple's Garden Retreat in Suzhou

It makes use of the elevation difference to stack Huangshi Rockery to form grotto scenes. A zigzag bridge was built to connect the cliffs at each side. It borrows scenes at Tinglu Building erected at the riverside, and is an excellent example to incorporate exterior scenes. This is an outstanding ravine design garden art in southern China.

苏州耦园入口与庭院组合
入口与庭院组合，雅趣顿生。

The Entrance of Suzhou Couple's Garden Retreat
Entrance combined with courtyard creats interest of elegancy.

苏州沧浪亭看山楼

建于印心石屋之上，前轩后楼，造型优美，昔日登轩可望苏州西南诸山。

Mountain View Building at the Canglang Pavilion

It was built on top of Yinxin Stone House, with a pavilion in front and building at the back. With its beautiful design, one could get on the top floor to enjoy distant mountains in southwest of Suzhou.

苏州寒山寺前庭 ｜ Front Courtyard of Hanshan Temple in Suzhou

苏州虎丘 | Tiger Hill in Suzhou

苏州鹤园馆房前假山景观 | Rockery Scene in Front of Crane Garden Building in Suzhou

苏州曲园曲池
在曲水亭下，池东回峰阁也暗含曲字，不但园“曲”，而且涵盖了园主人的一生。

Bend Pond at Bend Garden in Suzhou
Huifeng Pavilion at the east of the pond also suggests the word "Qu", meaning bending. Not only the garden is full of bending features, the whole life of the owner is alluded here to be anything but smooth or straight forward.

苏州拥翠山庄月驾轩
充分利用地形，假山结合磴阤，彰显兀立险峻。

Yuejia Pavilion at Yongcui Mountain Villa in Suzhou
Making full use of the terrain, its rockery is embedded with the climbing rock stairs, appearing to be steep and precipitous.

苏州同里镇退思园东部园林鸟瞰 | Bird's Eye View of Eastern Part of the Retreat & Reflection Garden at Tongli Town in Suzhou

苏州古镇木渎羡园

环山草庐楼前侧掇山，使楼与爬山廊皆融入山水环境，诗情画意油然而生。

Xian Garden at the Old Mudu Town in Suzhou
Rockery was built in front of Huanshan Caolu building, so that the building and the mount climbing corridor are well integrated into the landscape scene, creating a poetic and picturesque scene.

扬州瘦西湖精华——五亭桥、白塔、钓鱼台、凫庄

Essence of the Slender West Lake in Yangzhou: Five-Pavilion Bridge,White Pagoda, Fishing Terrace, Wild Duck Village

扬州瘦西湖五亭桥、白塔、凫庄鸟瞰 | Bird's Eye View of the Slender West Lake in Yangzhou: Five-Pavilion Bridge, the White Pagoda, and the Wild Duck Village

扬州瘦西湖熙春台、十字阁、落帆栈道、二十四桥（著者设计）

Xichun Terrace, Cross Pavilion, Luofanzhandao, Twenty-four Bridges at the Slender West Lake in Yangzhou (Designed by the Author)

扬州瘦西湖二十四桥景区设计效果图（著者设计）

Twenty-four Bridges Scenic Areas Design Rendering at the Slender West Lake in Yangzhou (Designed by the Author)

扬州瘦西湖熙春台（著者设计）
邻借望春楼，远借五亭桥、白塔，视野甚广阔。

Xichun Terrace at the Slender West Lake in Yangzhou (Designed by the Author)
It borrows the nearby Wangchun Building, and distant Five Pavilion Bridge and the White Pagoda, with very wide view angle.

扬州瘦西湖熙春台侧假山（著者设计）
不仅可登山道上楼，还体现了“叠石停云”的意境，使楼呈现腾云驾雾的艺术魅力。

Xichun Terrace Side Rockery at the Slender West Lake in Yangzhou(Designed by the Author)
It embodies a sense of climbing a hill and going upstairs, and also enables an artistic conception of "piled rocks to stop a cloud", making the building with a cloud treading artistic conception.

扬州瘦西湖“小李将军画本”
（著者设计）
借隔湖熙春台景收入扇面窗内，酷似一副装裱的山水画。

"General Li Jr. Album" at the Slender West Lake in Yangzhou
(Designed by the Author)
Borrowed by crossing Lake Xichun Terrace scene, the borrowed scene is contained in the fan-shaped window, just like a well-framed landscape painting.

扬州瘦西湖"卷石洞天"（著者设计）
峦崖洞壑与楼阁、廊桥结为一体，相得益彰，山顶四面八方亭供纵览诸景。

Juanshi Dongtian at the Slender West Lake in Yangzhou (Designed by the Author)
The cliffs and grottos integrated with the building, making them complement each other. The all-around pavilion on hill top offers a great spot to have an overview of surrounding scenes.

扬州瘦西湖“卷石洞天”庭院（著者设计）
“庭院以旱景水意”手法塑造石隙漫流的自然景观，配植芦苇加强水意。

Courtyard of Juanshi Dongtian at the Slender West Lake in Yangzhou (Designed by the Author)
Yard with imagined water scene forming imaginative random streams between rock cracks. Reeds are used to enhance the "water scene".

扬州瘦西湖“卷石洞天”的廊、桥、亭（著者设计）
起伏叠落的廊、桥、亭将全园分隔为水院和山庭两大区，各具特色。

Corridors, Bridges, and Pavilions of Juanshi Dongtian in Yangzhou (Designed by the Author)
They are at various heights divide the garden into water yard and rock yard two parts, with their own distinctive characteristics.

扬州瘦西湖“卷石洞天”薜萝水阁尺幅花窗（著者设计）
窗前为赏景佳处，观山、听瀑、赏花、览景，如入画境。

Chifu Flower Windows of Biluo Water Pavilion at Juanshi Dongtian at the Slender West Lake in Yangzhou (Designed by the Author)
A great spot to enjoy a scenery, one can view the mountain, listen to the sound of waterfalls, enjoy flowers and the scene, as if entering a picture.

扬州瘦西湖“白塔晴云”庭院框景（著者设计）

"White Pagoda Clear Cloud" Courtyard Framed Scene at the Slender West Lake in Yangzhou (Designed by the Author)

扬州瘦西湖“白塔晴云”前庭后院相互呼应（著者设计）

Front and Back Courtyard Corresponding to Each Other at "White Pagoda and Clear Cloud" at the Slender West Lake in Yangzhou (Designed by the Author)

扬州瘦西湖“白塔晴云”东庭院（著者设计）
三楹轩屋名“积翠”，园内湖石叠峰，青藤缠绕。南墙开水门引入湖水，乃私家游船出入与停泊的港湾。

East Courtyard of "White Pagoda Clear Cloud" at the Slender West Lake in Yangzhou (Designed by the Author)
The name of the building of Sanying Xuan is Jicui, namely the accumulation of greenness. In the garden there are stacks of rockeries and intertwined green vines. There is a water gate opening in the south wall to introduce lake water. This is a harbor for private pleasure boats going in and out.

扬州瘦西湖“白塔晴云”内庭半亭门厅（著者设计）
外墙水门使水系景色相互渗透、相得益彰。

Half Pavilion Entrance Hall at "White Pagoda Clear Cloud" Inner Courtyard at the Slender West Lake in Yangzhou (Designed by the Author)
The exterior garden wall's water gate makes water-oriented sceneries interlock and bring out the best in each other.

扬州瘦西湖“白塔晴云”
从回廊透视花南水北之堂
（著者设计）

Huanan Shuibei Hall Viewed from Zigzag Corridor Perspectively at Slender West Lake's "White Pagoda Clear Cloud" (Designed by the Author)

扬州瘦西湖“冶春诗社”
采用茅草屋面，别具韵味。

Yechun Poem Society at the Slender West Lake in Yangzhou
Thatch roof is used for its unique charm.

扬州瘦西湖“梅岭春深”是人工堆的全湖最高的山，有联句“如拳不大，金山也肯过江来”，故又名“小金山”。

"Deep Spring at Plum Ridge" at the Slender West Lake in Yangzhou
It is the highest rockery of the whole lake area. Because of a saying "if not as big as a fist, Mount Jin would come over from the other side of the river", it is also called "Little Mount Jin".

扬州瘦西湖“双峰云栈”（著者方案设计，江苏兴业环境集团参与施工）双峰云栈为恢复的历史景点。

"Shuangfeng Yunzhan" at the Slender West Lake in Yangzhou (Designed by the Author, and Jiangsu Xingye Environment Group Participated in the Construction)
This is a project for restoring a historical scene.

扬州瘦西湖静香书屋（著者设计）
书屋面对山水，一边与廊桥相接，另一边为院墙海棠门、院墙开水门，引水入院，景色曲折多变。

Jingxiang Study at the Slender West Lake in Yangzhou (Designed by the Author)
On one side it faces the rockery and the pond, and on the other a Chinese flowering crabapple shaped gate. Water is channeled into the courtyard through a water gate on the garden wall, with varied scenes.

扬州瘦西湖静香书屋主入口框景（著者设计）

Framed Scene at the Main Entrance of Jingxiang Study at the Slender West Lake in Yangzhou (Designed by the Author)

扬州瘦西湖静香书屋内庭院
（著者设计）
小中见大，巧依屋廊、半亭，并引水点石，门窗借景。

Inner Courtyard of Jingxiang Study at the Slender West Lake in Yangzhou
(Designed by the Author)
One envisions a larger landscape in this small and limited space. The yard cleverly leverages the half pavilion and the corridor, channeling water and dotting rocks and with borrowed scenes.

扬州瘦西湖静香书屋内东望天然桥与高低变化的曲廊
（著者设计）

Looking East towards a Natural Bridge and Meandering Corridor on Variant Height from Jingxiang Study at the Slender West Lake in Yangzhou (Designed by the Author)

扬州瘦西湖钓鱼台

三面皆有圆门，西圆门借景五亭桥卧波，南圆门借景白塔耸立十分巧妙。

Fishing Terrace at the Slender West Lake in Yangzhou

There are round doors in three sides, with its west one borrowing scenes of five-pavilion bridge over water, and south one white pagoda. Very ingenious.

扬州瘦西湖五亭桥上俯视凫庄
远借钓鱼台，小金山诸景。

Five Pavilions Bridge Borrowing Scenes from Wild Duck Village below at the Slender West Lake in Yangzhou
It borrows scenes of Fishing Terrace and Little Mount Jin etc. in the distance.

扬州瘦西湖群玉山房入门（著者设计）

有泉三叠，冷冷作响，声如琴韵，恍惚走进琼苑仙境。

Entrance of Qunyu Shanfang at the Slender West Lake in Yangzhou (Designed by the Author)

There is a three-cascade spring at the entrance with wonderful water dropping sound, as if one enters fairy land.

扬州瘦西湖吟月茶楼（著者设计） | Lingyue Tea House at the Slender West Lake in Yangzhou (Designed by the Author)

扬州瘦西湖徐园门景

门景为圆形，墙后为门房建筑，入园荷池映莲，石桥卧波。

Round Gate Scene at Xu Garden in the Slender West Lake in Yangzhou

Behind the wall is the gate house. After entering the garden gate, one sees stone bridges over lotus pond.

扬州瘦西湖扇形水榭

在柳岸深处，一湾碧水中，两侧游廊，东连方亭，西接六角亭，高低错落。相传乾隆皇帝南巡时，皇妃的梳妆台设于此处。

Fan-Shaped Water Pavilion in the Slender West Lake in Yangzhou

Located deep in willow shore, a clear water bay with corridors at both sides, connecting a square pavilion in the east and a hexagonal pavilion in the west at different heights. Legend says when Emperor Qianlong had his southern China tour, his imperial concubines used this place as their dressing spot.

扬州个园四季假山：“春”
石笋——春山淡冶而如笑。

Four Season Rockery "Spring" at A Garden in Yangzhou
Rock shoot: spring mountain light as a smile.

扬州个园四季假山："夏"
夏云——夏山苍翠而如滴。

Four Season Rockery "Summer" at A Garden in Yangzhou
Summer clouds: summer mountain is verdant and fresh.

扬州个园四季假山："秋"
登高——秋山明净而如妆。

Four Season Rockery "Autumn" at A Garden in Yangzhou
Ascending a height: autumn mountain bright and clean as if with a makeup.

扬州个园四季假山："冬"
积雪——冬山惨淡而如睡。

Four Season Rockery "Winter" at A Garden in Yangzhou
Accumulated snow: winter mountain is pale and sleepy.

扬州个园秋山上佛云亭景色 | Scene of the Buddha Cloud Pavilion on Autumn Hill of A Garden in Yangzhou

扬州天宁寺西园
（著者与江苏兴业环境集团风景园林规划设计院设计）即御花园复原效果图。

West Garden of Tianning Temple in Yangzhou
(Designed by the Author and Landscape Planning & Design Institute of Jiangsu Xingye Environment Group)
This is the restoration rendering of the imperial garden.

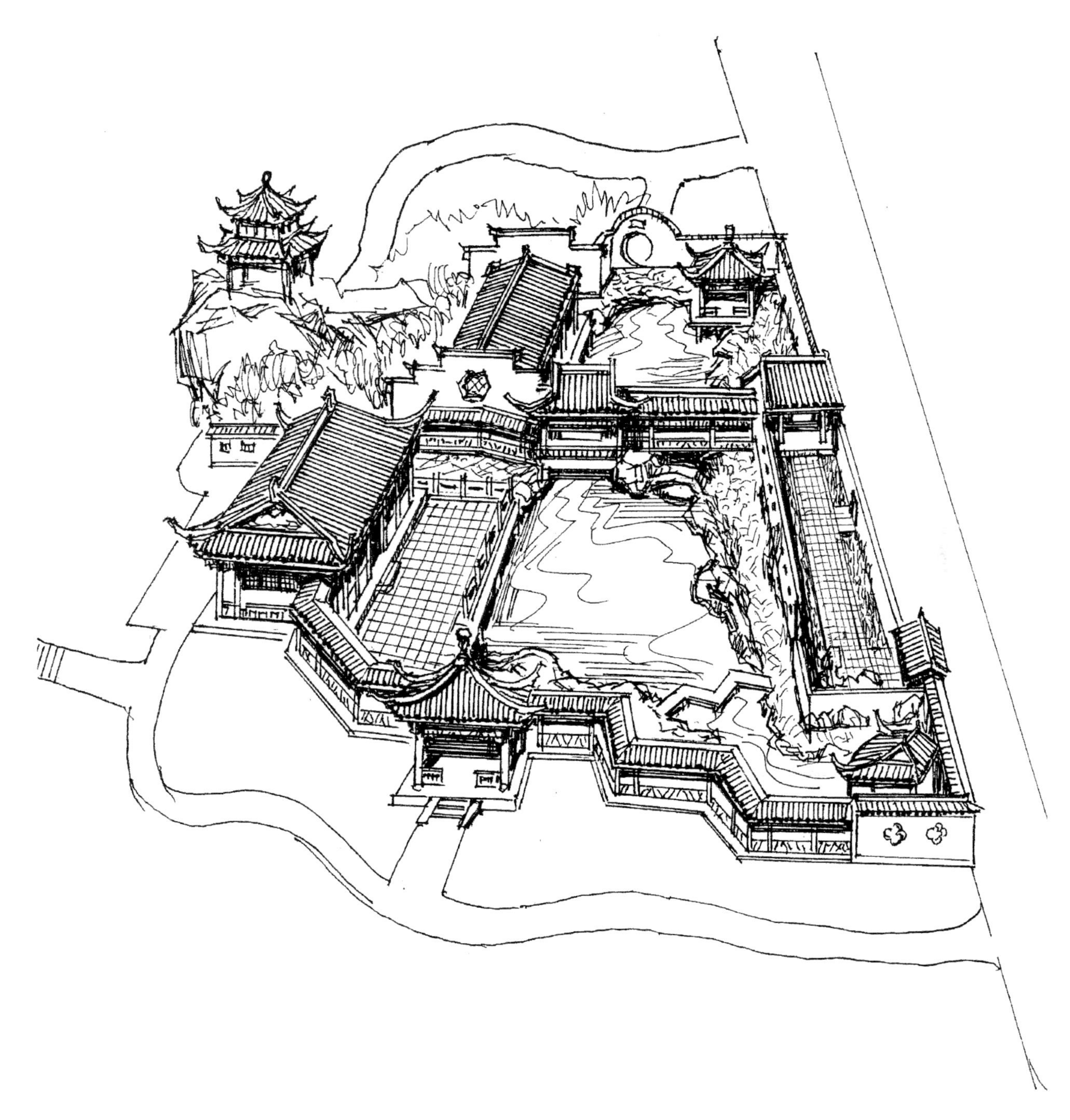

扬州天宁寺西康熙和乾隆皇帝南巡御宅复原图“云山胜地”（著者与江苏兴业环境集团风景园林规划设计院设计）

"Yunshan Shengdi", Restoration Rendering of the Royal Residence of Emperor Kangxi and QianLong's Southern China Tour, West of Tianning Temple in Yangzhou (Designed by the Author and Landscape Planning & Design Institute of Jiangsu Xingye Environment Group)

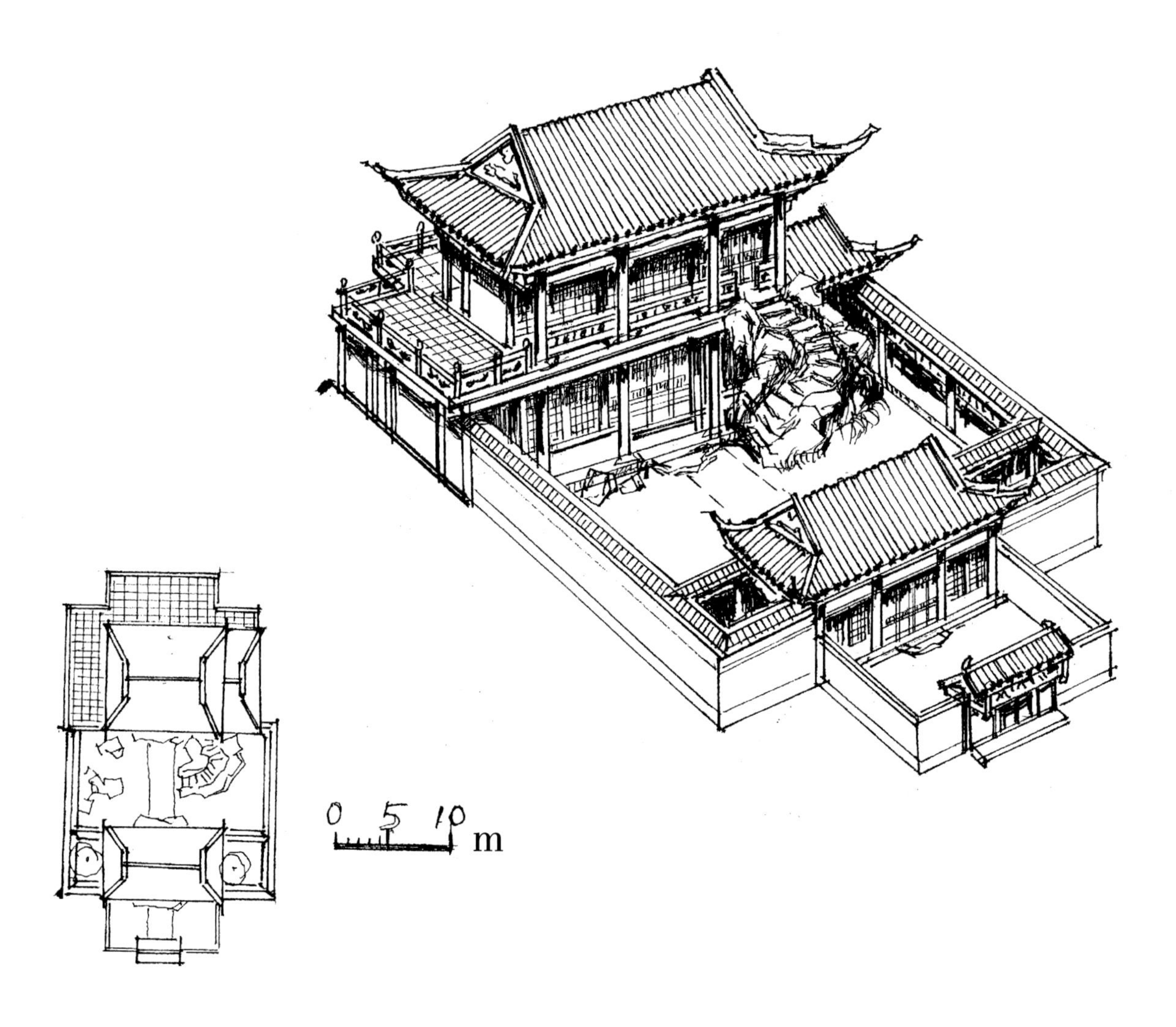

扬州天宁寺西康熙和乾隆皇帝南巡御宅复原图“静含太古山房”（著者与江苏兴业环境集团风景园林规划设计院设计）

"Jinghan Taigu Mountain House", Restoration Rendering of the Royal Residence of Emperor Kangxi and QianLong's Southern China Tour, West of Tianning Temple in Yangzhou (Designed by the Author and Landscape Planning & Design Institute of Jiangsu Xingye Environment Group)

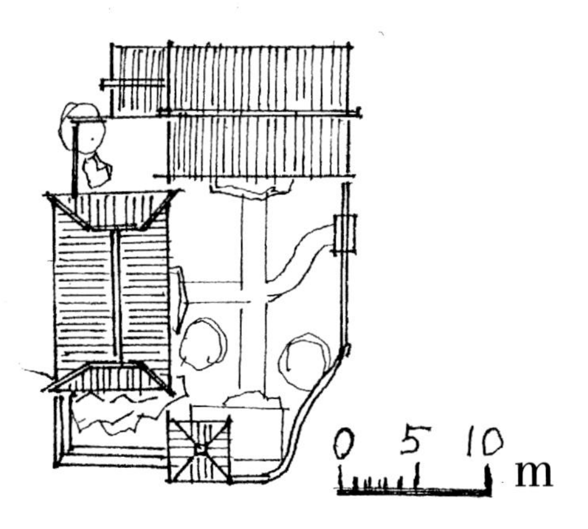

扬州寄啸山庄楼角处叠石
其造景汇合屋面雨水形成瀑布。

Overlaid Rockery at the Corner of Jixiao Mountain Villa Building in Yangzhou
The built rockery is actually a mechanism to collect rain water from the roof to form waterfalls.

扬州寄啸山庄月亭
是典型的园林之台。

Moon Pavilion in Jixiao Mountain Villa in Yangzhou
Moon Pavilion is a typical garden terrace.

扬州寄啸山庄水心亭
为池中方形凉亭，亦称“小方壶”，典雅古朴，倚栏赏景，可俯观游鱼之乐，仰视楼廊之美。

Pond Center Pavilion at Jixiao Mountain Villa in Yangzhou
It is a square pavilion in the pond, so it is also called "Little Square Kettle". Elegant with primitive simplicity, against the balustrades one can look down to enjoy the fish or look up to appreciate the beauty of architecture.

扬州片石山房（著者设计）
为明代大画师石涛手笔，故掇山酷似其画作，山高水长的造景恰似一副立体的国画长卷，在书房里眺望画意盎然。

Pianshi Shanfang in Yangzhou (Designed by the Author)
It is the work by the great artist Shi Tao in the Ming dynasty. His rockeries bear strong resemblance to his artistic paintings. The pursuit of "high as mountains and long as rivers" in artistic paintings can be realized here, as can be appreciated from inside the study.

扬州片石山房修复后效果
（著者设计）

Appearance after Restoration of Pianshi Shanfang in Yangzhou (Designed by the Author)

扬州“二分明月楼”日月形门窗（著者设计）
是园林艺术美的典范。

Sun and Moon Shaped Door and Window at "Erfen Bright Moon Building" in Yangzhou (Designed by the Author)
This a model of beauty in garden art.

扬州御赐文汇阁与园林复原设计
（著者与江苏兴业环境集团风景园林规划设计院设计）

Garden Restoration Design and Wenhui Pavilion, Bestowed by the Emperor, in Yangzhou (Designed by the Author and Landscape Planning & Design Institute of Jiangsu Xingye Environment Group)

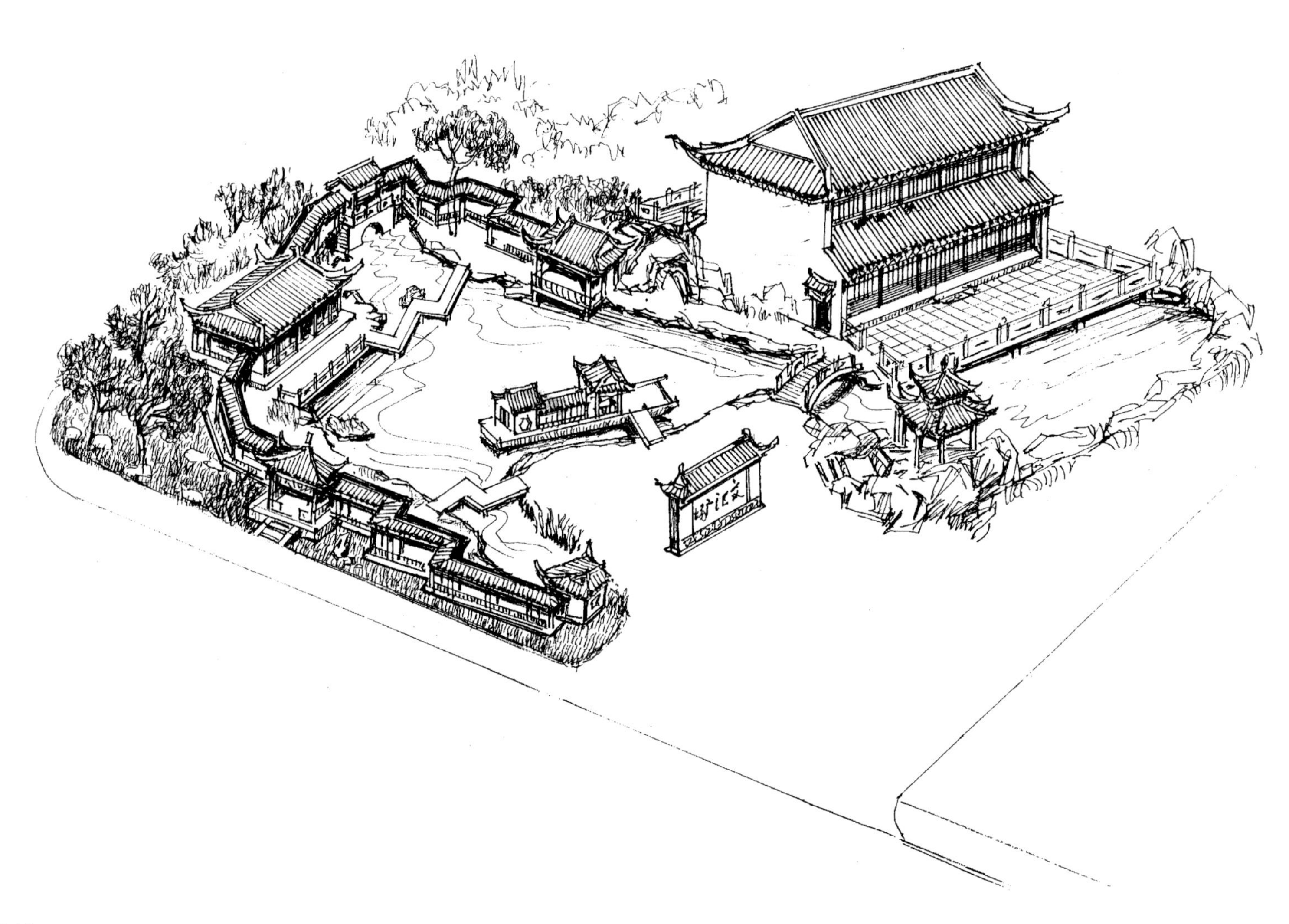

无锡寄畅园仰借锡山塔景
使园景层次极为丰富。

The Pagoda Scene Borrowed from Down Below by
Jichang Garden in Wuxi
It creates rich layers of landscape scenes.

无锡寄畅园磨砖雕花门楼
此乃江南古典园林最典型的做法。

Carved Brick Decoration Gate at Jichang Garden in Wuxi
This is the most typical practice in classical southern Chinese gardens.

无锡蠡湖春秋高阁
曲廊低回，流传着春秋时范蠡与西施的优美传说，让槛内人感到岁月的流逝，无情却有情。

Spring and Autumn High Pavilion at Li Lake in Wuxi
Legend goes about Fan Li and Xi Shi in ancient Chunqiu times in this zigzag veranda, making people here feel emotional about the pass of time while appreciating the scene.

无锡蠡湖宝界桥西南堍

山崖水际，矗立着八柱三门三楼二亭式琉璃顶牌楼，造型典雅，彩绘精美，是鼋头渚风景区的入口标志。

Southwest End of Baojie Bridge at Li Lake in Wuxi

At this joint place between the mountain and water, there erected an elaborate Pailou with colored glaze, which has eight columns, three gates, three stories, and two pavilions. With elegant form and exquisite color paintings, it is the entrance landmark of Yuantouzhu scenic area.

无锡太湖临水而筑的“涵万轩”
以半亭翼墙的简洁形式，让游人斜倚吴王靠，小憩探水，得景外之情。

Hanwan Pavilion Built by Waterfront at Tai Lake in Wuxi
Its form is half pavilion against a wing wall, by which visitors could lean against the King of Wu chair balustrade to explore the interests of nearby water and sceneries outside.

上海豫园湖心亭、九曲桥

Lake Center Pavilion and Zigzag Bridge at Yu Garden in Shanghai

上海豫园双层楼阁
屹立于池畔湖石假山之上，上层称快楼，下层名延爽阁，眺望园景如诗如画。

Double Story Building at Yu Garden in Shanghai
It is set on pond side lake rockeries, with Quick Building on top and Yanshuang Pavilion at the bottom. Looking over the garden one gets a picturesque scene.

上海豫园仰山堂与卷雨楼

在三穗堂后有一座窗饰华丽、飞檐高翘的二层楼厅，与大假山遥遥相对，下层为仰山堂，可仰望山景，俯赏游鱼，上层为卷雨楼，雨日登临，对岸烟雾迷蒙，发人遐想。

Yangshan Hall and Juanyu Building at Yu Garden in Shanghai

Behind Sansui Hall, there is a double-storied building with rich window decorations and high-flying eaves, facing the sizable rockery. Its base floor is called Yangshan Hall, where one can look up the mountain scene and look down to enjoy the swimming fish, while on the upper floor, on a rainy day, one may daydream at the other misty side.

上海秋霞圃霁霞阁

建在湖石堆叠的仙人洞上，北洞口有磴道至冈上，入阁从南洞口可步至晚香居。

Jixia Pavilion at Qiuxia Garden in Shanghai
Built on a rockery with a fairy grotto underneath, whose north entrance leads via steps to the top of the rockery. After entering Jixia Pavilion, the grotto's south entrance leads to Wanxiang Pavilion.

上海松江醉白池景观

三百多年来，醉白池历尽沧桑，但旧貌依稀，环池三面曲廊亭榭，以及池中山石堆叠，仍保持着明清江南园林的动人风姿。

Scenery at Zuibai Pond at Songjiang in Shanghai

Over three hundred years, this place experienced a lot, but still retained its appearance. Around the pond at three sides there are pavilions, halls, and covered zigzag corridors, as well as rockeries in the pond. It still retains its attractive charm of southern Chinese Ming and Qing dynasties' style.

杭州曲院风荷观景塔

Belvedere at Lotus in the Breeze at Crooked Courtyard in Hangzhou

杭州曲院风荷红绡翠盖廊
长廊掩映在浓荫深处，曲折多姿，满池碧叶红花，荷香沁人。

Red Thread Verdant Cover Corridor at Lotus in the Breeze at Crooked Courtyard in Hangzhou
The long corridor dwells deep in the thick shade, zigzagging its way with a fragrant full pond of red flowers and green leaves.

杭州西泠印社山顶庭院西视剖面图

West Section View of Mountain Top Courtyard at Xiling Seal Engraver's Society in Hangzhou

杭州西泠印社山顶园林北视图 | North View of hilltop garden at Xiling Seal Engraver's Society in Hangzhou

杭州郭庄

庄在卧龙桥北、金沙溪旁，濒湖筑台榭，以水池为中心，曲水与西湖相通，旁叠湖石假山，隔西里湖与苏堤相望。

Guo Village in Hangzhou

It is located north of Crouching Dragon Bridge by Jinsha Creek. Terraces and buildings were built by the lakeshore and they are centered around water. The water body is connected with West Lake. Rockeries are constructed at the building's side. Over the inner lake, Guo Village and Su Causeway look at each other.

杭州郭庄两宜轩

Liangyi Pavilion at Guo Village in Hangzhou

杭州郭庄凌波桥 | Lingbo Bridge of Guo Village at West Lake in Hangzhou

杭州芝园

中心为水池，桥亭如仙鹤飞来，飘然卓立于石桥之上，亭台楼阁皆成倒影，真可谓“汇景池”。

Zhi Garden in Hangzhou

Centered by a pond, the bridge and the pavilion are like a crane just arrived, flying over and perched on the balustraded bridge. Pavilions, terraces, buildings and halls all have their reflections in the water; thus it is called "Huijing Pond", or accumulated reflections pond.

杭州芝园延碧堂

意在把青山绿水美景引进来，厅前临水筑有露台，站在露台上，隔水池与大假山遥遥相对，一池碧水中，倒映楼台亭阁，是观赏美景的好地方。

Yanbi Hall at Zhi Garden in Hangzhou

It is intended to introduce outside lovely landscape scenes inside. This is a great site to enjoy the scene. In front of the hall by the water there is a terrace, where one could stand there to enjoy the large rockery over the pond. Building images are all reflected in the clear water pond.

杭州小瀛洲内湖东北九曲桥

贯通小瀛洲岛南北，共九转三回三十个弯。人行桥上有步移景异、曲径生情之感。

Northeast Zigzag Bridge in Inner Lake of Small Yingzhou in Hangzhou

It runs north-south wise through the island, and there are nine turns, three U-turns, and thirty bendings. On the pedestrian bridge, one gets a novel view at every change of direction. Great interests arise on turning garden paths.

杭州虎跑泉庭院南视剖面图 | Section South View of Running Tiger Spring Courtyard in Hangzhou

杭州御风楼
由于高居假山之顶，是昔日杭州城南最高建筑。

Yufeng Building in Hangzhou
As it is perched high on a rockery, it was the highest building south of the Hangzhou city.

南京瞻园南假山
临池屹立，有绝壁悬崖、峰峦、洞龛、山谷、瀑布、石矶、汀步等，为假山中的精品。

South Rockery in Zhan Garden in Nanjing
It stands just by the pond, with precipice cliffs, rock peaks and grottos, canyons and waterfalls, rocky ledges and stepping stones etc. It's one of the finest works on rockeries.

南京瞻园西部一水抱三山佳境

东部有庭院，水院组成清逸古朴的景观。图为院中部景色，三十二曲回廊将两景区连缀成一体，堪称佳绝。

Beautiful Scene Showing Three Mounts Embraced by Single Water Body in Western Part of Zhan Garden in Nanjing

There is a courtyard in the east. The water yard creates a scene of primitive simplicity. This shows the mid-section scene of the garden. It uses a thirty-two turns ambulatory corridor to connect the two scenic areas. It's an excellent example.

南京瞻园三十二曲回廊

为园东、园西联景、隔景的重要建筑物，构制奇巧，千姿百态，相形度势，宛若长龙。

Thirty-two Winding Corridor at Zhan Garden in Nanjing

This is a connecting scene between the eastern and western gardens. The important buildings for separating scenes was built with ingenious techniques using various forms. It accommodates concrete situations and appears as a long dragon.

南京瞻园延晖亭

为园中夕阳余晖最终驻足之处，与叠落廊相接，景致前伸后延。

Yanhui Pavilion of Zhan Garden in Nanjing

It is the spot in the garden where the afterglow of a sunset finally sets. It connects with the overlapping corridor and its scene shows front and hind extensions.

南京莫愁湖赏荷厅

两面夹水，有掬水之便，得月之光，赏荷之雅，纳凉之爽。登厅西阁上四方亭，远可尽览湖山之胜，近可观荷池中神情娴静的莫愁女石雕像。

Lotus Appreciation Hall at Mochou Lake in Nanjing

The building is sandwiched between water bodies and has the convenience of scooping up water nearby, getting the moonlight, and enjoying the lotus' elegance. Getting onto the square pavilion above the west building, one can have an overview of the mountains and lakes, and at close distance can enjoy the statue of calm and demure Lady Mochou in the pond.

北京琼岛春阴 | Qiongdao Chunyin in Beijing

绍兴兰亭曲水流觞
一弯流水蜿蜒，九曲回环，为当年兰亭修禊觞咏之地。

Orchid Pavilion's Qushui Liushang in Shaoxing
Meandering flowing creek is where versifying happened in ancient time.

嘉兴南湖鱼乐园

其碑刻系明万历三十年董其昌手书，鱼池湖石嶙峋，游鱼嬉戏，此乃典型金鱼缸的章法。

Fish Pleasure Garden at South Lake in Jiaxing

Its inscriptions show Dong Qichang's handwriting calligraphy. With jagged lake rocks and frolicking fish, this is the typical goldfish-tank way of design.

淮安衙署后花园主体景观（著者设计）

Back Garden Main Scenery of an Government Building in Huai'an (Designed by the Author)

宁波神采卓然天一阁

秦氏支祠尤为华丽，支祠戏台、回廊、木雕构件，无不流光溢彩、熠熠生辉，堪称甬上建筑之精华。

Magnificent Tianyi Pavilion in Ningbo

The Qin family branch ancestral hall is a magnificent building. Its stage, corridors, wood building components, all display rich shining colors. It can represent the elite class in Yongshang architecture.

河南鲁山泉佛寺鸟瞰图（著者设计）

Bird's Eye View of Spring Buddha Temple at Mount Lu in Henan (Designed by the Author)

河南鲁山泉佛寺山门正立面图（著者设计）

The Front Facade of the Main Gate of Spring Budda Temple (Designed by the Author)

临沂琅琊园北湖亭、廊、桥、榭景色（著者设计）

Scenery of Pavilion, Corridor, Bridge, and Terrace at the North Lake of Langya Garden in Linyi (Designed by the Author)

临沂琅琊园水榭与廊亭，并移植古树陪衬（著者设计）

Water Terraces and Corridor Pavilions Accompanied by Transplanted Old Trees at Langya Garden in Linyi (Designed by the Author)

临沂琅琊园楼榭（著者设计）
以廊相连，环抱水池以弧形桥分隔成大小水面，鱼龟分养。

Buildings and Pavilions at Langya Garden in Linyi (Designed by the Author)
They are connected by a corridor, which embraces the pond. A curved bridge divides the water body, making two ponds, one for keeping fish and the other turtles.

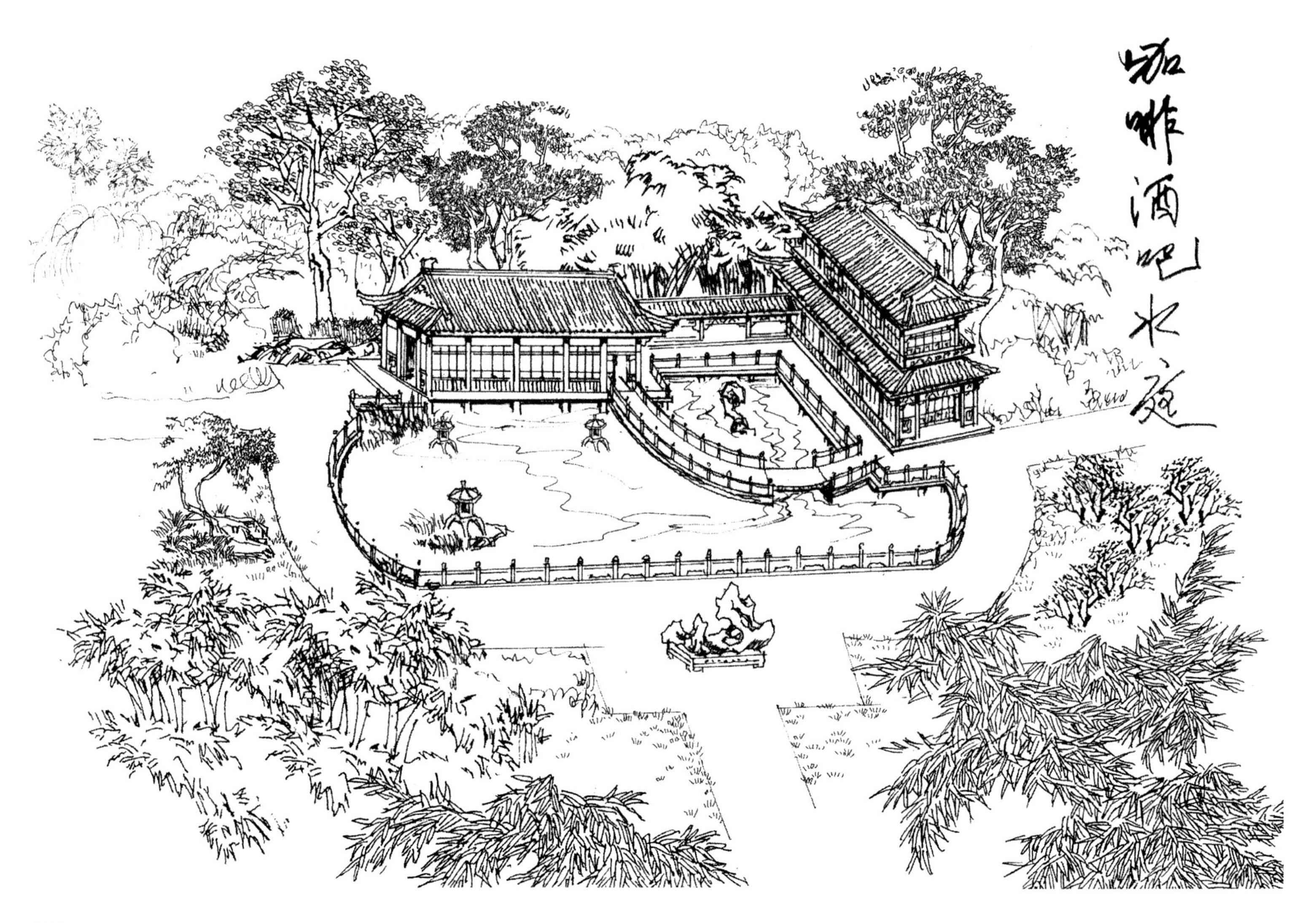

临沂琅琊园主景——建在山上的琅琊阁（著者设计）

Langya Pavilion, the Main Scenic Spot Built on Top of a Hillock in Langya Garden in Linyi (Designed by the Author)

临沂琅琊园湖边的不系舟（石舫）（著者设计）
寓意个性解放。

The Unbound Boat (Made of Stones) at the Lakeshore of Langya Garden in Linyi (Designed by the Author)
The boat represents liberation of individual character.

临沂琅琊园——松风里苑（著者设计）
盆景园之门是用湖石拼接的拱门，并与围墙组合而成，别具意趣，匠心独运，此乃国内之首创，亦是盆景大师范义成的佳作。

Songfengli Park of Langya Garden in Linyi (Designed by the Author)
This bonsai garden's entrance gate is an arch made of lake rocks and it is integrated with the garden wall. This interesting, charming, and unique garden gate is creative in China, and it is also a great work of bonsai master Fan Yicheng.

临沂琅琊园——待月井亭（著者设计）
岩崖临水，特建水亭，而亭中水井与亭顶圆孔相结合，待中秋佳节月影投入井中，此乃仙境也。

Moon Waiting Well Pavilion of Langya Garden in Linyi (Designed by the Author)
The water pavilion was built by the rock cliff. The well in the pavilion corresponds with the round hole on the roof. At Mid-autumn Festival the moon casts its reflection into the well, creating a fairyland atmosphere.

临沂琅琊园——琅琊云阁（著者设计）
云阁参照古代界画仙境建筑皆是十字脊攒尖的法式，而山体有画论“一峰突起，连岗断堑”“山欲动而势长”的神韵。

Langya Cloud Pavilion of Langya Garden in Linyi (Designed by the Author)
The building was built according to ancient Chinese fairyland architecture building style required cross-based pyramidal roof. The hill is based on fine art painting theory of "one outstanding peak with broken chasm but connecting range" and "mountain poised to move with power".

临沂琅琊园——沂河莲渡（著者设计）
园内水体引自沂河，从莲花桥下涌入，继而萦回，体现“四边水色茫无际”的诗情画意。

Yi River Lotus Ferry of Langya Garden in Linyi (Designed by the Author)
The water in the garden is from the Yi River, and it is introduced from under the Lotus Bridge. Its goal is to construct a lingering effect to realize "water water everywhere" poetic atmosphere.

临沂琅琊园——琅琊汉门（著者设计）

园门是历史的展现，同时融入文化与艺术，激起游客参观的欲望。

Langyahan Gate of Langya Garden in Linyi (Designed by the Author)

The garden gate is not only an exhibition of the history, it is also blended in art and culture, to whet the appetite for people to visit.

临沂琅琊园——新甫苍柏

古柏原生于新甫山下，今移入园中，树龄已超过2000余年，仍焕发苍劲古朴的美感，傲骨铮铮的性格续写琅琊新篇。

Xinfu Old and Strong Cypress of Langya Garden in Linyi

This ancient cypress originally grew at the foothill of Mount Xinfu, and now it has been transplanted into the garden. It is already over two thousand years old; however it still thrives and exhibits its vigorous and primitive beauty of simplicity, and continues to extend its pride.

临沂琅琊园——承露天台
（著者参加设计）
古人凌晨于天台上承接露水，经山岩石隙漫流至涧溪，形成五叠涌泉，场面壮观。

Dew Collecting Sky Terrace of Langya Garden in Linyi (Designed by the Author)
People in ancient times collected dew at early morning. The water then flowed down through rock crevices and reached bottom canyon creeks, forming five cascade gushing springs, very spectacular.

临沂琅琊园——书圣崖瀑（著者参加设计）
瀑布从山坳涌出，分三级跌落，下为深潭，后流入涧溪，可谓“得自然之理，而呈自然之趣”，故画意盎然。

Calligraphic Prodigy Cliff Waterfall of Langya Garden in Linyi (Designed by the Author)
The waterfall gushes out from cols, flows down in three cascades, falls into deep pond below, and finally flows into mountain streams. This is so called "appearing naturally interesting after getting nature's rules"; therefore picturesque.

翠明园进入园门第一组庭院（著者设计） | First Series of Courtyards After Entrance at Cuiming Garden (Designed by the Author)

翠明园延山引水，画意盎然
（著者设计）

Cuiming Garden Extending Mountain Ranges and Channeling Waters, Forming Picturesque Scenes (Designed by the Author)

翠明园院墙转角（著者设计） | Courtyard Wall Corner at Cuiming Garden (Designed by the Author)
半亭贴壁，画意甚佳。 | The half pavilion is attached to the wall, creating great artistic inspirations.

四季庭院设计——“春”· 春笋
（著者设计）

Four Season Courtyard Design — "Spring" ·
Spring Bamboo Shoot (Designed by the Author)

四季庭院设计——“夏”· 荷池（著者设计）

Four Season Courtyard Design — "Summer" · Lotus Pond (Designed by the Author)

四季庭院设计——“秋”· 登高（著者设计）| Four Season Courtyard Design — "Autumn" · Ascending a Height (Designed by the Author)

四季庭院设计——“冬”· 风雪（著者设计）

Four Season Courtyard Design — "Winter" · Wind and Snow (Designed by the Author)

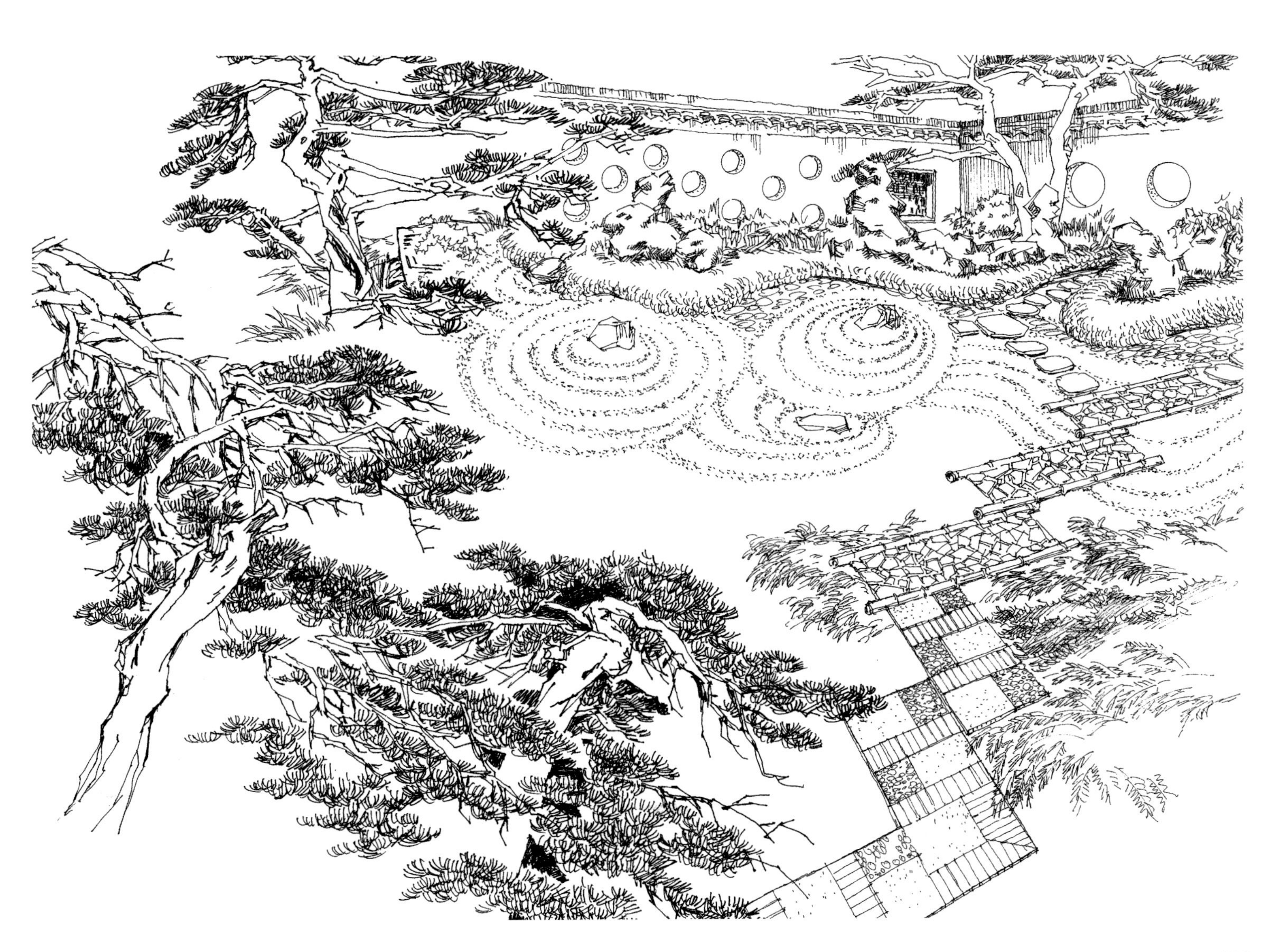

戏台与观众排座（著者设计）| Stage and Audience Seat Arrangement (Designed by the Author)

丝竹琴台（著者设计）
“行到水穷处，坐看云起时。”

Sizhu Qintai (Designed by the Author)
"Traveling to where the water ends, sitting till the cloud rises."

莲坛禅韵（著者设计）
“掬水月在手，弄花香满衣。”

Liantan Chanyun (Designed by the Author)
"The moon is in your hands when you hold water reflecting it, you are immersed in fragrance when you handle flowers."

烟渚柔波（著者设计）
“岛影含树影，天光透水光。”

Yanzhu Roubo (Designed by the Author)
"Tree images in island's reflections, sparkling water shimmering with skylight."

岩泉鹿嬉（著者设计） | Yanquan Luxi (Designed by the Author)

一朝风月（著者设计）
“延山引泉谋禅境，亭桥廊榭营唐风。”

Yizhao Fengyue (Designed by the Author)
"Zen is induced when extending mountains and introducing springs; Tang dynasty atmosphere is re-created with pavilions, bridges, corridors and trees."

沁园云境（著者设计）
“月落潭无影，云生山有衣。”

Qinyuan Yunjing (Designed by the Author)
"No shadow of the moon in the pond, the mountain obtains a clothing of cloud shadow cast."

苇舟野渡（著者设计）
“始从芳草去，又逐落花回。”

Reed Boat at the Wild Ferry (Designed by the Author)
"Leaving with fragrant grass, returning following falling flower petals."

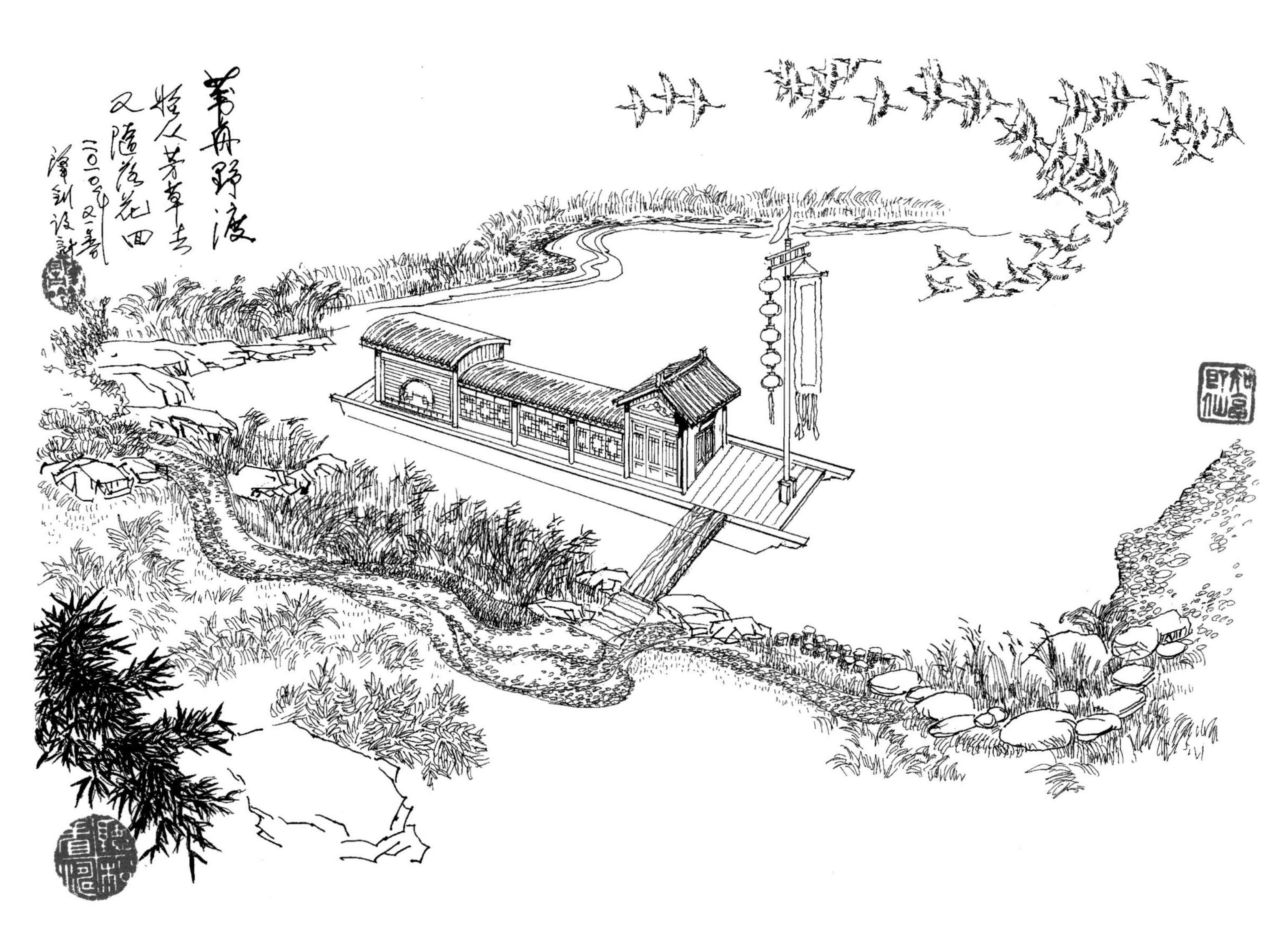

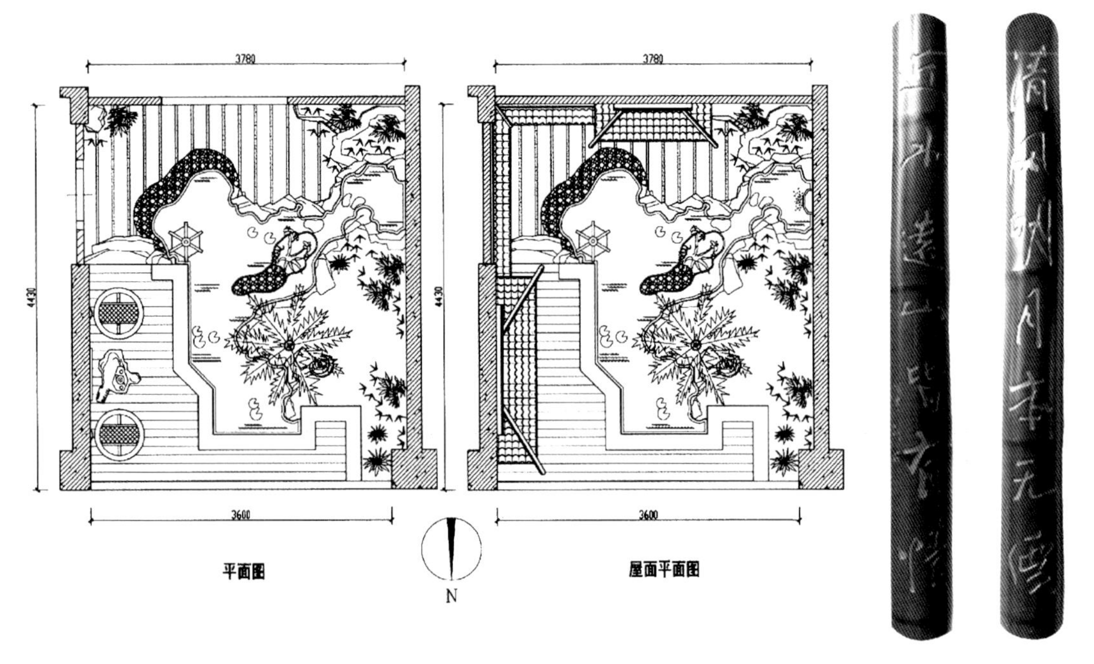

高层住宅亦可享受庭院
（著者设计）

One Can also Enjoy a Yard Garden even in High-rise Residences (Designed by the Author)

江南园林开水门

方便乘船，可直通市肆水路，这也是水乡造园的特色。

Open Water Gate in Southern Chinese Gardens

It is so designed for the convenience of boat boarding and easy access to the market via waterways. This is also characteristic in garden design in regions rich in rivers and lakes.

四面八方“非必丝与竹，山水有清音”（著者设计）

"Enjoy the Scenic, Not Necessarily with Music" at All Directions (Designed by the Author)

蝉噪林愈静，鹊鸣山更幽（著者设计） | Quieter Woods with Cicada Noise, Secluded Hillock with Bird Chirping (Designed by the Author)

清风明月本无价，近水远山皆有情（著者设计）

Clear Wind and Bright Moon are Priceless, Love Fills Water and Mountain Everywhere (Designed by the Author)

错落左右林野趣，横斜高低石铿锵
（著者设计）

Interest in Wild Forests and Interlacing Rocks
(Designed by the Author)

古建别墅的门亭、围墙设计（著者设计）

Entrance Pavilion and Garden Fence Wall Design in Old-Styled Villas (Designed by the Author)

人造假山展览馆（著者设计） | Artificial Rockery Exhibition Building (Designed by the Author)

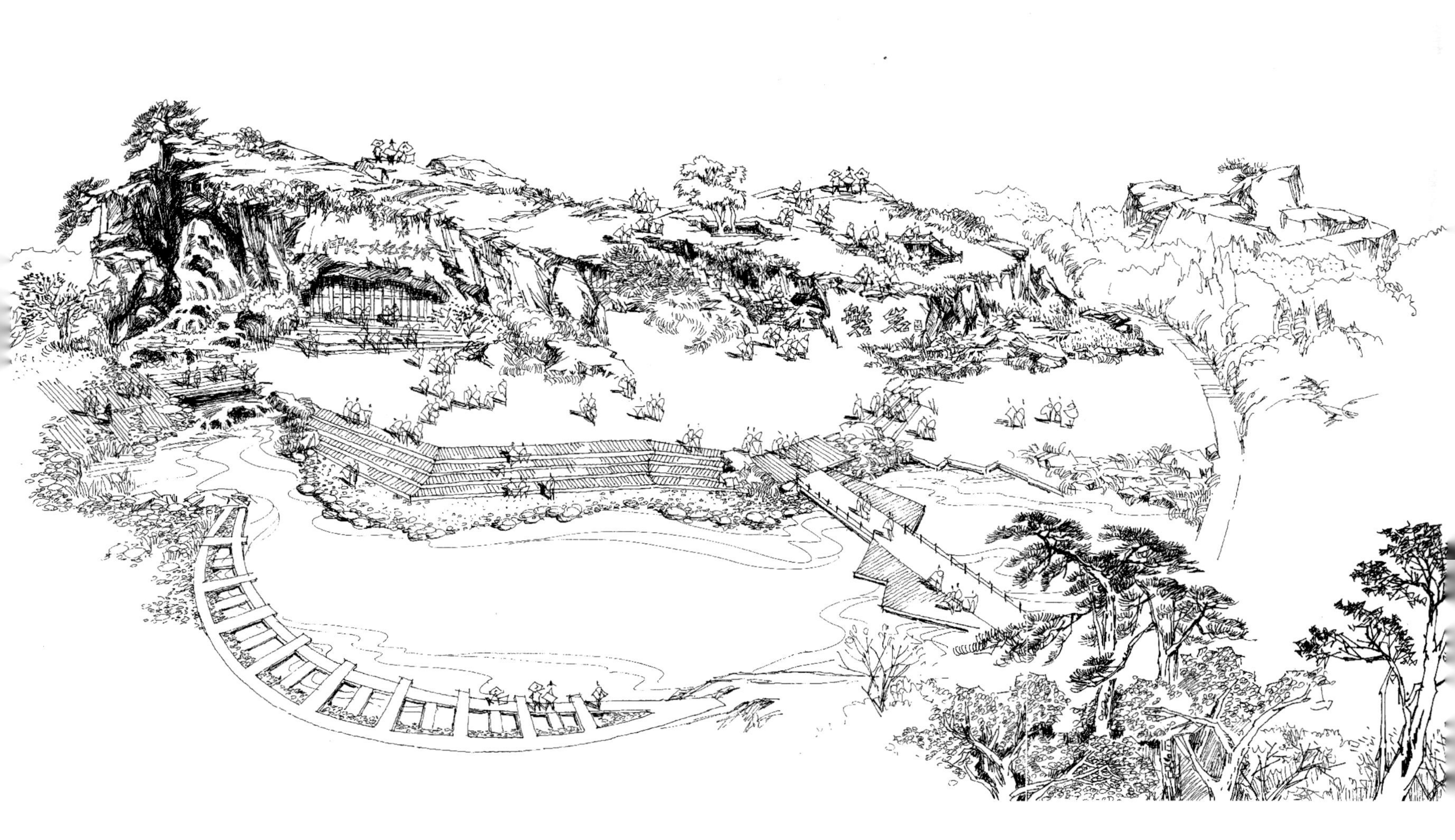

江南水乡水上歌舞表演台 | Stage for Singing and Dancing in Southern China Water Town

参照岭南民居围屋设计的客家餐饮与娱乐会所
（著者设计）

Hakka Catering Club Design Based on Lingnan Circularly Enclosed Residence (Designed by the Author)

突出延山引泉造园手法的景致（著者设计）
石隙漫流可谓以假乱真。

Scene Stressing on Extending the Mountain and Channeling the Stream (Designed by the Author)
Streams flowing in rock crevices makes one think it is real.

禅境小游园创意图（著者设计） | Zen State Small Garden Conception (Designed by the Author)

园西部曲溪潺流
有桃花源意境，水尽头为活泼的水阁。

Meandering Creek in Western Part of the Garden
It has the artistic perception of Taohuayuan. At the end of the water body is lively pavilion built above water.

碧驾玉庭江南园林鸟瞰图
（著者与江苏兴业环境集团风景园林规划设计院设计）

Bijia Yuting Bird's Eye View of Southern Chinese Garden (Designed by the Author and Landscape Planning & Design Institute of Jiangsu Xingye Environment Group)

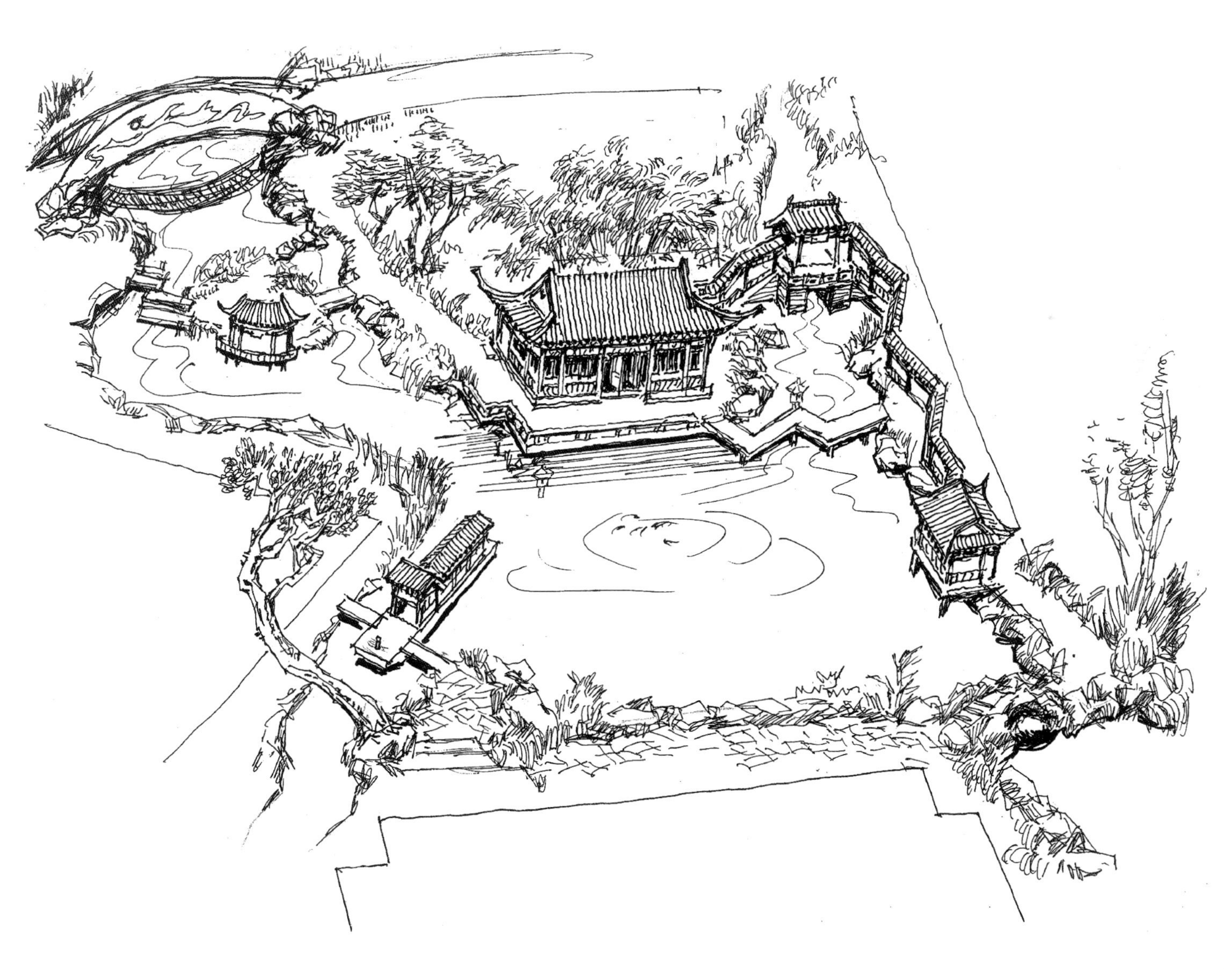

江南园林景点设计“琴台”
（著者与江苏兴业环境集团风景园林规划设计院设计）
“流水蜿蜒，九曲四环。”

"Musical Instrument Terrace", Scenic Spot Design of Southern Chinese Garden (Designed by the Author and Landscape Planning & Design Institute of Jiangsu Xingye Environment Group)
"Water is bending and twisting around."

江南园林景点设计“画舫”
（著者与江苏兴业环境集团风景园林规划设计院设计）
“行到碧驾处，坐看玉庭时。”

"Artistic Painting Boat", Scenic Spot Design of Southern Chinese Garden
(Designed by the Author and Landscape Planning & Design Institute of Jiangsu Xingye Environment Group)
"Travel to where the green boat is, when it is time to enjoy the jade yard."

江南园林景点设计“棋轩”
（著者与江苏兴业环境集团风景园林规划设计院设计）
“岛影含树影，天光透水光。”

"Ancient Chess Playing Pavilion", Scenic Spot Design of Southern Chinese Garden
(Designed by the Author and Landscape Planning & Design Institute of Jiangsu Xingye Environment Group)
"Reflections of the island contain that of the trees, light of water passes through the sky light. "

江南园林景点设计“扇面亭”
（著者与江苏兴业环境集团风景园林规划设计院设计）

"Fan-Shaped Pavilion", Scenic Spot Design of Southern Chinese Garden (Designed by the Author and Landscape Planning & Design Institute of Jiangsu Xingye Environment Group)

理石系列

C

ROCKERY ARRANGEMENT SERIES

苏州羡园云墙之下
水门贯通二园内外盘曲的涧水，故题“槃涧”。

Under the Cloud Wall at Xian Garden in Suzhou
The watergate connects the inside and outside meandering creek water bodies, thus so-called "coiled ravine".

苏州狮子林庭园叠石小品“牛吃蟹”
似显幽默。

"A Cow Is Eating a Crab" Rockery Feature at the Lion Grove Garden Courtyard in Suzhou
It appears to be humorous.

苏州狮子林探幽海棠形洞门框景九狮峰 | Chinese Flowering Crabapple Shaped Door Frame Scene Nine Lions Rock, Prompting for Exploration at the Lion Grove Garden in Suzhou

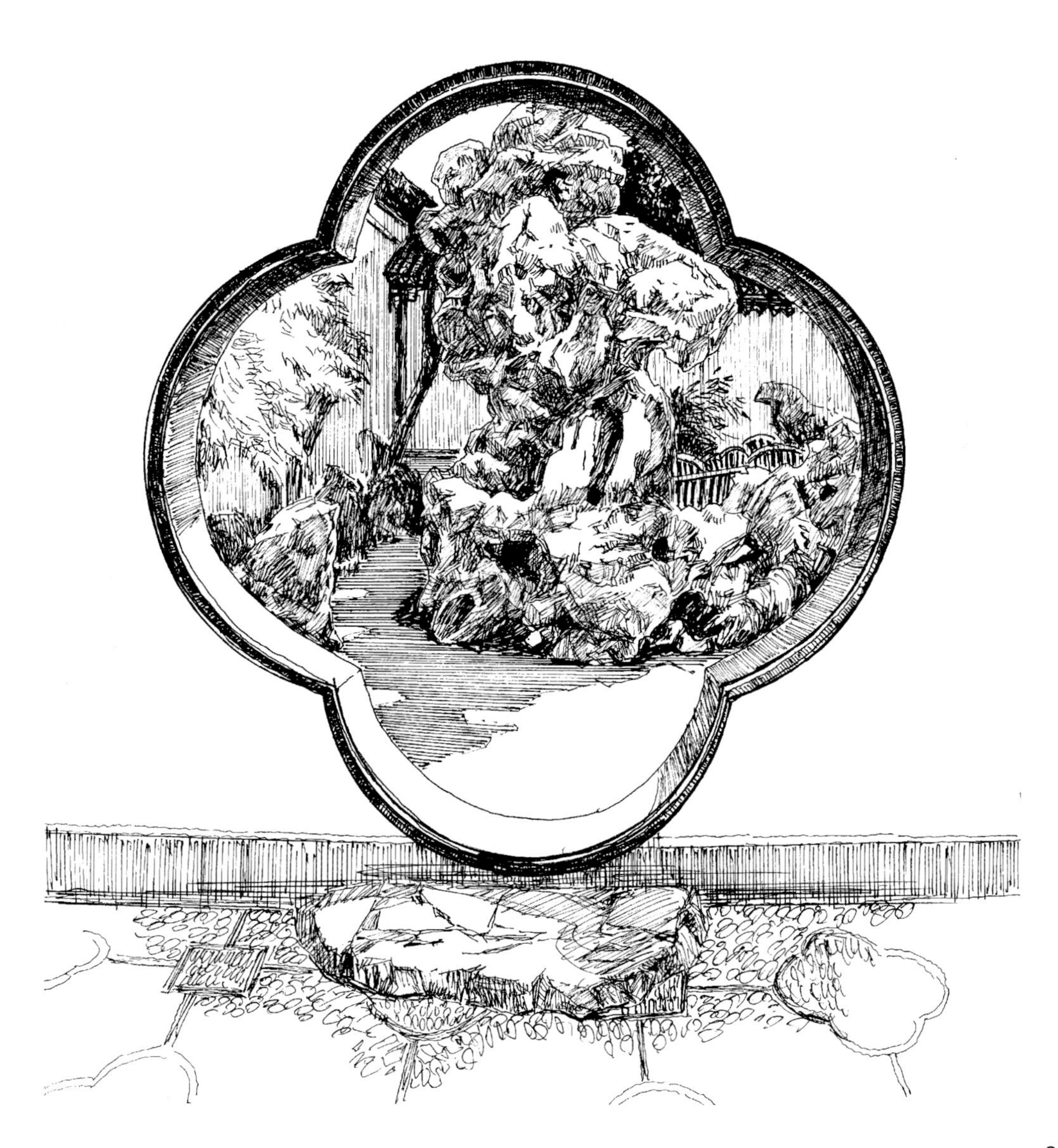

苏州留园揖峰轩处
峰石林立，故名石林小院。花台中栽植牡丹，寓意花开富贵。

Jifeng Pavilion at the Lingering Garden in Suzhou
Groves of rock peaks contribute to the name of the small rock grove yard. Peonies are planted in flower beds to implying blossom and riches.

苏州留园华步小筑 | Huabu Miniature Features at the Lingering Garden in Suzhou

苏州拙政园海堂春坞 | Haitang Chunwu at the Humble Administrator's Garden in Suzhou

苏州网师园梯云室框景 | Framed Scene at Tiyun Room at the Net Master's Garden in Suzhou

苏州瑞云峰

Ruiyun Rock in Suzhou

苏州冠云峰

Guanyun Rock in Suzhou

苏州屏风三叠 | Three-Folded Screens in Suzhou

苏州九狮峰
在小方厅北，是一组堆叠巧妙的湖石，峰峦有若狮形，妙在似与不似之间。

Nine Lions Rock in Suzhou
This is a group of ingeniously composed lake rocks. The peaks and ridges resemble a lion, and the wonder lies between likeness and not likeness.

上海古猗园五老峰
是散置于逸野堂右侧的五座石峰，因状似聆听山水清音的老人而得名。

The Five-Old-Man Rock at Old Yi Garden in Shanghai
The five rock peaks are scattered on the right hand side of Yiye Hall. It was named after the image of old men listening to the music of nature.

上海豫园跨水游廊中方亭
竖有立峰，名“美人腰”，是进入大假山的前奏。

Square Pavilion at Cross Water Covered Corridor at Yu Garden in Shanghai
Inside the pavilion there is a standing rock, which is called "Beauty Waist", and it is the prelude to entering large rockeries.

曲折有致，错落得体 | Well-Arranged Twists and Turns

盘曲溪流，野趣横生 | Winding Streams with Wild Interest

贝聿铭设计的北京香山饭店庭院中的瀑布 | Waterfall in Xiangshan Hotel Designed by Ieoh Ming Pei in Beijing

芜湖翠明园庭院角隅的瀑布（著者设计）

Courtyard Corner Waterfall at Cuiming Garden in Wuhu (Designed by the Author)

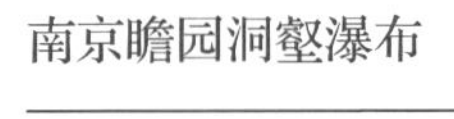
南京瞻园洞壑瀑布

Grotte Waterfall at Zhan Garden in Nanjing

南京瞻园假山、石矶汀步 | Rockery and Stepping Stones at Zhan Garden in Nanjing

南京莫愁山
堆叠于池畔，池中山影似幻化倩影。

Mount Mochou in Nanjing
Stacked at the pond side, the reflection of rockeries in the pond is shifting and illusive.

苏州虎丘天平山 | The Mount Tianping at Tiger Hill in Suzhou

盂泉
用毛竹造成二级跌落，水盂为自然山石，周围铺装卵石、白沙。

Jar Spring
Using mao bamboo to create two cascades. The water collection container is made of raw rocks, and its surroundings are pebbles and white sand.

盂泉与水中的石灯笼相映成趣 | Jar Spring and Rock Lantern in the Pond Interestingly Contrasting Each Other

日本人造规则式瀑布景观 | Japanese Formal Style Artificial Waterfall Scene

扬州瘦西湖卷石洞天景区群玉山房（著者设计）

内室角隅点缀湖石洞龛、毛竹引泉，使室内生机勃勃，“山水有清音”诗句跃然于画面。

Clustered Jade Mountain Villa in Juanshi Dongtian Scenic Area at Slender West Lake in Yangzhou (Designed by the Author)

Using mao bamboo to channel spring water makes a lively interior environment, reminding visitors of the verse "music abounds in natural landscapes".

扬州片石山房石涛画幅以实景再现（著者设计）| Shitao's Painting Realized at Pianshi Mountain Villa in Yangzhou (Designed by the Author)

扬州江都仙女公园“清音”泉流（著者设计） | "Qingyin" Streams at Fairy Park of Jiangdu in Yangzhou (Designed by the Author)

厦门公园小品 | Garden Features in Xiamen Public Park

广州勐园塑石景观佳作 | Fine Sculpted Rockery Work at Meng Garden in Guangzhou

广州宾馆内庭佳作“故乡水”

Inner Court Fine Work "Hometown Water" at Guangzhou Hotel

广州白云宾馆为保存古树而塑造的景观

Scenery for Preserving Ancient Trees at Baiyun Hotel in Guangzhou

广州花园酒店塑石佳品 | Fine Sculpted Rockery at Garden Hotel in Guangzhou

广州兰圃水口小品 | Water Outlet Feature at Orchid Garden in Guangzhou

扬州瘦西湖望春楼内室（著者设计）
采用“室内天井”的格局，楼上下可共享假山景观，室内有山麓的野趣，满楼烟云的意境是江南园林中的孤例，可谓匠心独运。

Inner Room of Wangchun Building at Slender West Lake in Yangzhou (Designed by the Author)
A patio is employed in the interior, so that a rockery scene can be shared by both the ground and the upper floors. Wild mountain effect is created in the interior, plus the whole house foggy atmosphere, it is the only example in southern Chinese gardens, very ingenious.

风景区生态景观
（著者设计）

Ecological Scenery in Scenic Area
(Designed by the Author)

李清照词园“木簟花”景区（著者设计）
以李清照词“花自飘零水自流”句为创作依据。

"Mudianhua" Scenic Area at Li Qingzhao Poem Garden (Designed by the Author)
It is based on Li Qingzhao's poem "flowers scatters on free flowing water".

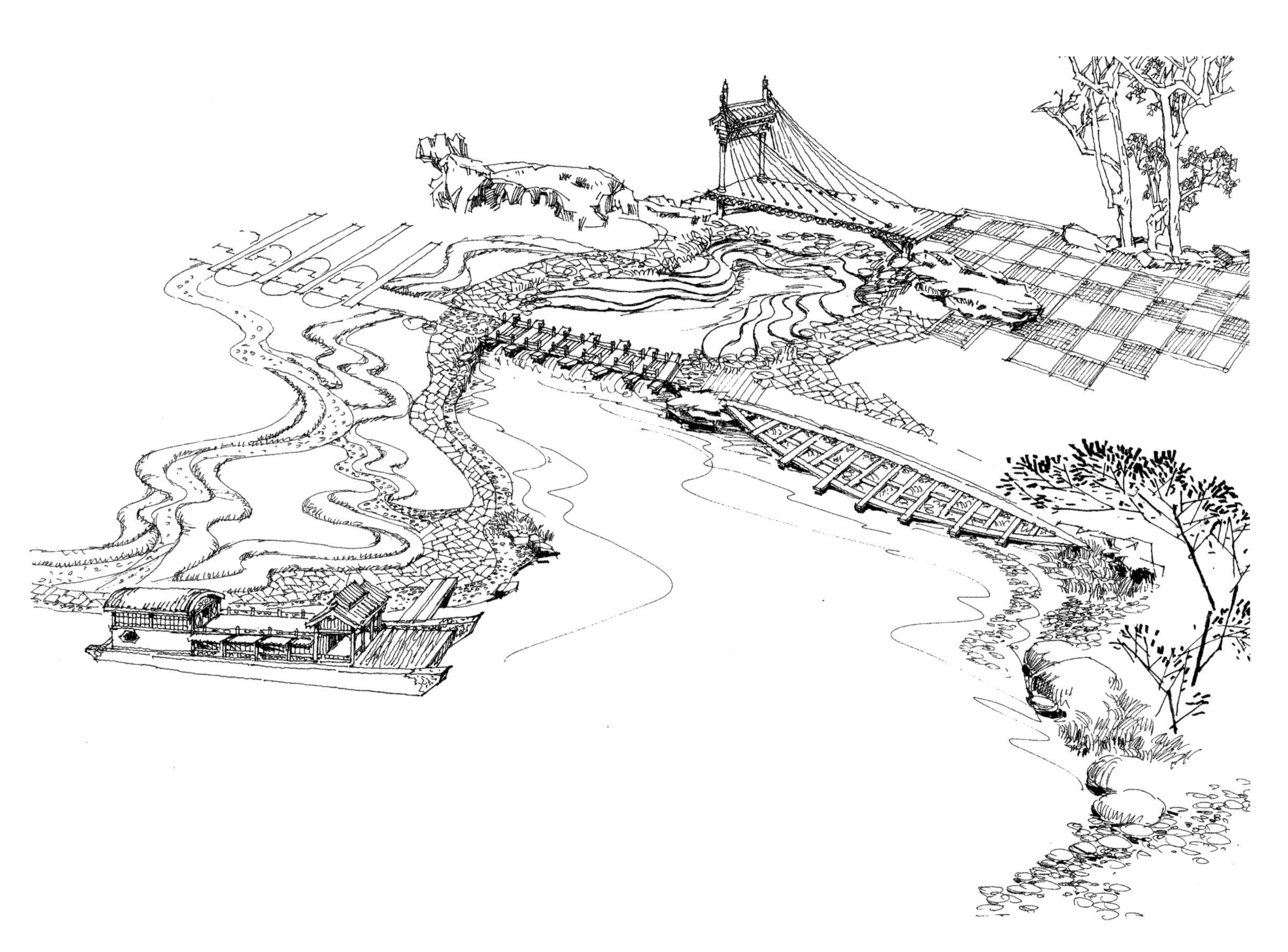

水洞深邃
（著者设计）

Deep Water Tunnel
(Designed by the Author)

无锡二泉池前明代太湖石峰“观音石”

以粉墙为纸，藤石为绘，有八大山人的写意画境。

Ming Dynasty "Goddess of Mercy Rock" Tai Lake Rockery in Front of the Second Spring in Wuxi

It uses the white painted wall as the background, and vine-climbed rocks as the subject, and manifests itself as a freehand brushwork artistic painting of the famed Badashanren.

苏州怡园岁寒草庐南庭院山石小品

Rockery Features in the South Courtyard of Suihan Caolu at Yi Garden in Suzhou

苏州天下第三泉
框景

Framed Scene at
the Third Spring
in Suzhou

山石使带状水系产生深远的艺术效果

Rockeries Creating a Sense of Depth Artistic Effects in Linear Pattern Water Bodies

常州兴福寺涌泉“空心潭”

Gushing Spring "Hollow Pond" of Xingfu Temple in Changzhou

苏州网师园冷泉亭
内特置英石峰，别具韵味。

Cold Spring Pavilion of Net Master's Garden in Suzhou
Inside quartz rock enjoys its unique charm.

苏州怡园入口石景 | Entrance Rockery Scene of Yi Garden in Suzhou

可通画舫的洞府与洞壁叠泉
（著者与孟兆祯院士合作设计）

Grottos that Lead to Artistic Painting Boat and Grotto Side Cascade Springs (Designed by the Author and Academician Meng Zhaozhen)

临沂琅琊园涌泉待月亭（著者设计） | Gushing Spring Daiyue Pavilion of Langya Garden in Linyi (Designed by the Author)

扬州二十四桥景区朝西的望春楼（著者设计）
利用建筑外廊营造内室叠石，开南门引入瘦西湖景色，得到楼内外景观融为一体的效果。

West Facing Wangchun Building of Twenty-four Bridges Scenic Areas in Yangzhou (Designed by the Author)
It leverages the building's exterior corridor to arrange interior rockery. The opening of the south door introduces scenes of Slender West Lake. Therefore it creates an integration effect blending both the interior and the exterior scenes.

苏州留园绿荫轩
由“华步小筑”和“古木交柯”二组小院环抱，简洁大气。

Verdant Shade Pavilion of the Lingering Garden in Suzhou
The two small yards of "Huabu Xiaozhu" and "Gumu Jiaoke" embrace each other, simple yet magnanimous.

紫竹苔院

"苔痕映阶绿，修篁入帘青。"

Purple Bamboo Moss Yard

"Traces of moss against greenery, slender bamboos into curtained verdancy."

镇江金山
利用塔影湖东北云根岛山石险峻，建风月亭登临观景。

Mount Jin in Zhenjiang
Leveraging the steep mountain rocks of Yungen Island, Which is in the northeast of the Taying Pagoda, Fengyue Pavilion was built for people to enjoy the scene.

高层住宅亦可享受庭院（著者设计）
坐书房桌对景抽烟、聊天、看书、品茶、看鱼。

One Can also Enjoy a Courtyard in Highrise Buildings (Designed by the Author)
The scene facing the desk in the study, smoking, chatting, reading, tea tasting, and fish watching.

德国芳华园壁泉 | Wall Spring at Fanghua Garden in Germany

海外系列

D

OVERSEA
SERIES

美国纽约大都会博物馆中国园“明轩”

Ming Pavilion in Chinese Garden at New York City Metropolitan Museum in the US

美国纽约大都会艺术博物馆中国园“明轩”冷泉亭 | Cold Spring Pavilion in Ming Pavilion of Chinese Garden at New York City Metropolitan Museum of Art in the USA

加拿大温哥华逸园 | Yi Garden in Vancouver, Canada

澳大利亚悉尼谊园 | Friendship Garden in Sidney, Australia

澳大利亚悉尼达令港谊园之船舫 | Boat Building at Friendship Garden in Darling Harbour, Sidney, Australia

德国清音园门亭 | Gate Pavilion of Qingyin Garden in Germany
（著者设计） | (Designed by the Author)

德国斯图加特清音园思谊厅（著者设计） | Siyi Hall in Qingyin Garden in Stuttgart, Germany (Designed by the Author)

德国郢趣园 | Yingqu Garden in Germany

德国法兰克福春华园 ｜ Chunhua Garden in Frankfurt, Germany

德国斯图加特清音园四面八方亭
（著者设计）

All Around Pavilion of Qingyin Garden in Stuttgart, Germany (Designed by the Author)

德国幗园 | Guo Garden in Germany

德国法兰克福春华园 | Chunhua Garden in Frankfurt, Germany

德国慕尼黑芳华园 | Fanghua Garden in Munich, Germany

英国利物浦燕秀园 | Yanxiu Garden in Liverpool, the United Kingdom

中国香港墓园景观 | Cemetery Garden Scene in Hong Kong, China

中国香港某家族墓园标志亭（著者设计）

Landmark Pavilion of a Family Cemetery Garden in Hong Kong, China (Designed by the Author)

日本大阪世博会同乐园（一）｜ Tongle Garden at the World Expo in Osaka, Japan (I)

日本大阪世博会同乐园（二）｜ Tongle Garden at the World Expo in Osaka, Japan (II)

新加坡蕴秀园（一） | Yunxiu Garden in Singapore (I)

新加坡蕴秀园（二）｜ Yunxiu Garden in Singapore (II)

泰国曼谷智乐园相映亭 | Xiangying Pavilion of Zhile Garden in Bangkok, Thailand

埃及开罗秀华园 | Xiuhua Garden in Cairo, Egypt

海外中国园方案（一）
（著者设计）

Oversea Chinese Garden Design (I)
(Designed by the Author)

海外中国园方案（二）
（著者设计）

Oversea Chinese Garden Design (II)
(Designed by the Author)

莲花亭组
（著者设计）

Lotus Pavilion Cluster
(Designed by the Author)

多种古建筑形式的组合（著者设计）| Combinations of Various Forms of Ancient Architecture (Designed by the Author)

参考文献 / REFERENCES

1. 陈植．园冶注释．北京：中国建筑工业出版社，1988．
2. 刘敦桢．苏州古典园林．北京：中国建筑工业出版社，1979．
3. 潘谷西．江南理景艺术．南京：东南大学出版社，2011．
4. 陈从周．江南园林．上海：上海科学技术出版社，1983．
5. 江苏省志．风景园林志．南京：江苏古籍出版社，2000．
6. 江苏省基本建设委员会．江苏园林名胜．南京：江苏科学技术出版社，1982．
7. 罗哲文，陈从周．苏州古典园林．苏州：古吴轩出版社，1999．
8. 杨瑾，薛德震．中国园林之旅．石家庄：河北教育出版社，2006．
9. 孟兆祯．孟兆祯文集．天津：天津大学出版社，2011．
10. 杨鸿勋．江南园林论．上海：上海人民出版社，1994．
11. 彭一刚．中国古典园林分析．北京：中国建筑工业出版社，1986．
12. 瘦西湖．瘦西湖风景区编印．
13. 陈从周．中国厅堂・江南篇．上海：上海画报出版社，2003．
14. 文震亨．长物志．北京：中国建筑工业出版社，2010．
15. 李渔．一家言．北京：中国建筑工业出版社，2010．
16. 朱有玠．岁月留痕：朱有玠文集．北京：中国建筑工业出版社．2010．
17. 吴肇钊．园冶图释．中国建筑工业出版社．2012．

跋

／

POSTSCRIPT

吴肇钊教授1966年毕业于北京林学院园林系，毕业后在扬州古典园林建设有限公司、中外园林建设有限公司、深圳市中外园林建设有限公司、深圳大学等单位工作。四十多年来，他主持的园林设计与施工工程达200余项；代表国家参加德国举行的“1993国际景园建筑博览会”，设计和施工的“清音园”荣获国际“大金奖”，并获德国政府荣誉奖章；是著名的园林设计专家，原建设部、广东省、深圳市政府专家组专家顾问，退休后任江苏兴业环境集团风景园林规划设计院总顾问。

黑发积霜织日月,教书育人显担当。吴教授十分注重对江苏兴业环境集团风景园林规划设计院新一代设计师的培养，带他们考察现场，给他们开讲座，“手把手”“一对一”地教，精雕细琢，润物无声；同时带领设计团队完成了一系列设计作品，得到了业界同仁的高度赞誉和褒奖，使设计师们快速地成长起来。在团队培养和教学的过程中，设计师们将吴教授的钢笔手绘作品悉心收集，进行整理、分类，汇编著成《江南园林艺术手绘图志》(江苏兴业环境集团风景园林规划设计院指定教材)一书。该书全面展示了他的广博学识、深厚功底，以及匠心独运的设计作品和令人瞩目的成果。该书集学术性、艺术性、欣赏性、实用性于一体，可供相关人员学习和参考。

在该书付梓之际，我代表江苏兴业环境集团向吴教授对我公司做出的贡献及对设计师们的培养表示感谢！对该书的正式出版表示祝贺！

江苏兴业环境集团有限公司董事长

二〇二一年三月十八日

附：江苏兴业环境集团有限公司简介

江苏兴业环境集团有限公司（以下简称“公司”）位于江苏省扬州市，公司注册资本1亿元。是一家民营多元化集团企业，是国家非物质文化遗产代表性项目“传统造园技艺（扬州园林营造技艺）”传承保护单位之一。公司业务范围涉及古典建筑、园林绿化、市政、建筑等多个行业，拥有市政、建筑、古建筑、装修装饰、（原）城市园林绿化共5个一级资质，城市及道路照明工程施工二级资质，水利水电工程、钢结构工程、环保工程、文物保护施工、公路工程、路基工程、路面工程、桥梁工程、隧道工程共9个三级资质，以及风景园林设计甲级资质。公司是江苏省扬州市唯一拥有园林设计甲级和古典建筑、（原）园林绿化双一级资质及园林EPC总承包资格的企业。获得资质的数量和层次在市内同行业企业中名列前茅，在省内同行业民营企业中屈指可数

公司在市委、市政府的领导下，在相关部门和单位的关心支持下，沿着“市外—省外—海外”的发展轨迹，一步一个脚印，积极践行扬州园林“走出去”的理念。公司先后承建和参建了雄安新区悦容公园中苑一区施工总承包项目、苏州工业园区体育中心、扬州智谷科技综合体工程、双峰云栈、蜀冈樱花大道、三湾湿地公园、宋夹城体育休闲公园、文昌西路绿化提升、第十届江苏省园博会扬州园、盐城斗龙港生态组团城市森林公园、安徽颍上五里湖湿地公园、抚顺月牙岛生态公园等多个标志性项目，海南文昌南洋美丽汇仿古建筑工程设计与施工一体化、宿迁酒都文化公园总承包EPC项目、京杭大运河（广陵段）沿河绿化提升和环境整治EPC项目、扬州维扬社区景观工程设计施工一体化项目、扬州史可法路景观提升设计施工一体化项目，第十一届中国（郑州）国际园林博览会“扬州园”设计项目、江苏旅游职业学院景观提升设计项目等，还在美国洛杉矶参与建设了“内润园”等项目。

2018、2019、2020年公司连续被评为“AAA级资信企业”，连续获得“江苏省建筑业百强企业”“扬州市文明单位”“扬州市建筑业先进企业”荣誉。公司获“鲁班奖”2项，

中国风景园林学会“园林工程奖”银奖2项、铜奖1项，“科学技术奖”三等奖3项，江苏省“扬子杯”优质工程13项，市级优质工程20项。公司所建项目的质量与品质受到了相关方面和社会的一致认可。

公司负责人多次荣获“优秀企业家”和“优秀企业经理”称号，2020年胡正勤董事长被评为江苏省第二批乡土人才“三带”名人、扬州市古建园林名师工作室领衔专家、邗城乡土名匠（名师）。

目前公司拥有各类工程技术人员和管理人员180人，其中教授2人，有高级工程师职称者13人，占8.3%；有工程师职称者50人，占27.8%；技术员45人，占25%；专业结构（构成）合理。

公司在胡正勤董事长的带领下，积极响应江苏省“积极探索体制机制创新，推动扬州工艺、扬州园林等传统技艺‘走出去’，拓展发展空间”。的指示和落实扬州市委“举全市之力推动扬州园林园艺走出去”的战略目标和任务。始终秉承“致力城市建设、服务社会发展”理念，积极实施优质品牌战略，加强人才引进和加大科技投入。以良好的社会信誉和综合实力努力创造更多的优质、精品工程，适应新形势、开拓新平台、实现新跨越、践行“中国梦”。